Small Farmer Production Systems: Way Forward

Small Farmer Production Systems: Way Forward

Editors:

Prof S Ayyappan, NABARD Chair Professor
ICAR-NDRI, Southern Regional Station
Adugodi, Bengaluru 560030

Dr Letha Devi G, Senior Scientist
ICAR-NIANP
Adugodi, Bengaluru 560030

Dr S Subash, Scientist
ICAR-NDRI, Southern Regional Station
Adugodi, Bengaluru 560030

Dr MCA Devi, Senior Scientist
ICAR-NDRI, Southern Regional Station
Adugodi, Bengaluru 560030

Dr P K Dixit, Principal Scientist
ICAR-NDRI, Southern Regional Station
Adugodi, Bengaluru 560030

SSPH

SATISH SERIAL PUBLISHING HOUSE

403, Express Tower, Commercial Complex
Azadpur, Delhi-110033 (India)
Phone : 011-27672852, Fax : 91-11-27672046
e-mail : info@satishserial.com, hkjain1975@yahoo.com
Website : www.satishserial.com

Published by :

SATISH SERIAL PUBLISHING HOUSE

403, Express Tower, Commercial Complex, Azadpur, Delhi-110033 (INDIA)
Phone : 011-27672852 Fax : 91-11-27672046
E-mail : info@satishserial.com, hkjain1975@yahoo.com

ISBN 978-93-86200-25-9

Composed, Designed & Printed in India

FOREWORD

The small farmer production system is an inevitable and integral part of Indian agriculture. The small and marginal farmers account for more than 85% of farm holdings. While declining farm size, coupled with occupational migration is a concern with regard to farming sustainability, the inherent advantages of smallholder farming in terms of transactional costs and family labour cannot be overemphasised.

Categorization of farms according to the scale of operation, incorporating all facets of agriculture is important for formulating appropriate policies for each section of the farming community. In a diversified agriculture canvass of the country, it is necessary to consider not just the land holdings, but the agro-ecosystem such as irrigated or rainfed; farm practices of foodgrains, cash crops, livestock, aquaculture and so on. Further, the success or the failure of small farms is determined strongly by the policy environment and access of farmers to inputs and information.

Increasing fragmentation of land, natural resource degradation, enhancing resource-use efficiency, access to quality inputs, enhanced energy usage, post-harvest management, transformation from farmer to agri-preneur, effective technology dissemination, linking farmers with markets, and enabling policy framework are the aspects to be considered for enhancing small farmers' income and welfare. Accordingly, there is a need to redesign the approaches for R&D and policy support in agriculture to ensure assured profitability and sustainability of small farmers.

I am happy to know that a National Consultation on 'Small Farmer Production Systems-Way Forward' was organised at Bengaluru, during 22-23 December, 2016, with experts in the field. I would also like to compliment Dr. S. Ayyappan, NABARD Chair Professor and his colleagues, for bringing out this publication as an outcome of the Consultation and wish this would pave way for addressing the issues of Smallholder farmers with a new approach.

15th June, 2017

Harsh Kumar Bhanwala
Chairman, NABARD

PREFACE

Agriculture in Asia is characterized by smallholders, with 87% of the world's 500 million small farms, likely to account for more than 91% by 2030. Small farmers are characterized by smaller applications of capital, but higher use of labour and other family-owned inputs, and a generally higher index of cropping intensity and diversification. While there are inherent disadvantages in terms of small size of farms, access to inputs, appropriate skills and poor terms of trade, the family farms grow a wide variety of cultivars, many of which are landraces with greater resilience against vulnerability to various stresses. However, they have low marketable surplus and profit. To ensure livelihood security of the marginal and small farmers, it is necessary to focus on the technological needs and infrastructure, including diversifying avenues for gainful employment in the non-farm sector.

Further, Agriculture is the largest private nano-enterprise in India, dominated by small farmers, who cultivate 44% of land, but contribute over 50% of total farm output. Their diverse activities contribute to food basket, nutritional security and household income of farmers and play a significant role in generating gainful employment in rural areas, particularly among the landless, small and marginal farmers and women. Marginal, small and semi-medium operational holders (area less than 4 ha) own about 87.7% of livestock and hence, development of livestock sector would be more inclusive.

The smallholder agriculture thus required to be analysed and approached with a different mind set and yardstick, to make it remunerative and sustainable. In order to analyse the role of small farmers in production systems, characterise the group dynamics of small farmer production systems and formulate interventions for doubling income of small farmer production systems, a National Consultation on 'Small Farmer Production Systems-Way Forward' was organised at Bengaluru, during 22-23 December, 2016, with experts in the field and this publication is an outcome of the Consultation.

We are grateful to Dr. H.K. Bhanwala, Chairman, NABARD, Mumbai; Dr. H.R. Dave, Deputy Managing Director, NABARD, Mumbai; Dr. M.V. Ashok, Chief General Manager, NABARD, Mumbai; Dr. H. Shivanna, Vice Chancellor, UAS (B), Bengaluru; Dr. A.K. Srivastava, Member, ASRB, New Delhi & Former Director, ICAR-NDRI, Karnal; Dr.

Raghavendra Bhatta, Director, ICAR- NIANP, Bengaluru; Dr. Sreenath Dixit, Director, ICAR-ATARI, Bengaluru; Dr. K.P. Ramesha, Head, ICAR-NDRI SRS, Bengaluru; and NAAS Secretariat, New Delhi, for the guidance and facilitations of the National Consultation. Thanks are due to the authors for enriching this publication with their intellectual inputs. We are hopeful that this publication would be a forerunner in Smallholder farming in the country for times to come.

15th June, 2017 **Editors**

CONTENTS

Chapter-1

Sustainable development of small farmers

K. NARAYANA GOWDA[1], K.C. NARAYANA SWAMY[2] AND N.S. SHIVALINGE GOWDA[3]

[1] Former Vice Chancellor, [2] Professor (Sericulture), Professor (Extension) [3]
University of Agricultural Sciences (B), Bengaluru

Largest number of farm families in India is constituted by small farmers. number of these families is bound to increase in coming decades in view of divisions and fragmentations of land holdings. These holdings are becoming uneconomical which is further aggravated due to fragmentation of holdings and majority of these families are located in rainfed areas. Small farmers are generally facing problems of low financial resources, do not own pair of draught animals to take care of timely land preparation, sowing andother farm operations. They are low risk bearing farmers in view of low confidence and limited knowledge. Due to aforesaid background great majority of them did not adopt improved technology as in case of large farmers. Many studies have pointed out that either they have gone for delayed adoption or no adoption and in some cases partial adoption. In this process, gap between rich farmers and poor farmers has widened. This was very much evident particularly after Green Revolution in India. In order to narrow gap between resource rich farmers and resource poor farmers, many client based programmes were introduced for small farmers development during 1970's. Most important ones are:

1. Small Farmer's Development Agency, 1970-71
2. Marginal Farmer's and Agricultural Labourer's Programme, 1970-71
3. Integrated Rural Development Programme-1978
4. Lab to Land Programme Sponsored by ICAR, 1979

5. District Rural Development Agency/Society, 1980
6. Some states have also started specialized programmes like 'Anthyodaya' by then Rajasthan government to improve socioeconomic conditions of these families.

Most of these programmes were implemented in selected parts of country, they were time bound, poor linkages and more emphasis on physical changes than behavioural changes without addressing end to end issues, particularly marketing. As a result, improvement brought about due to these programmes did not sustain in long run.

In recent years, dependency on external inputs and support has increased to all types of farmers, more so for farm power with introduction of tractorization while in yester years, dependency was minimal and villages were self contained to a greater extent. Draught animals were owned by almost all families in yester years and gradually declined particularly among small farmers. Conditions of small farmers in recent years particularly after commercialization and globalisation of a agriculture has become very weak both on resources and their confidence to continue in farming. Most majorities of large farmers have shifted to high value crops and gone for commercialisation of agriculture. standard of living of large farmers has improved considerably while small farmers' living conditions remained to be same or further declined. Dependency of small farmers on large farmers has increased on timely farm operations besides certain critical inputs, non formal sources of loan and certain socio phychological factors in social system. Both State and Central Governments have launched several programmes after 1990's including Institution Village linkage programme (IVLP) by ICAR, but they did not address wholistic development as well as end to end issues. Keeping all these developments, UAS, Bangalore took initiative to bring in sustainable development of farmers in general, and small farmers in particular through implementation of Rural Bio resource Complex project.

Rural Bioresource Complex Project (RBRC)

RBRC project funded by Department of Biotechnology, Government of India was implemented by the University of Agricultural Sciences, Bangalore from April 2005 to March 2010 for a period of five years in Tubagere Hobli of Doddaballapura Taluk, Bangalore Rural district, Karnataka state. Major objective was to bring improved standard of living of farmers including landless families and ensure farming as a profitable venture on a sustainable basis. Baseline survey was conducted to know existing scenario and new opportunities for development of these families. There were 8340 families spread over in 75 villages which included eight percent large farmers, 66 percent small farmers and 26 percent landlessfamilies.

Further analysis on distribution of farm families in project area revealed that 89.12 percentage of total farm families were small farmers. Development of small farmers cannot be addressed in isolation without giving due attention to other category of farmers in view of negative attitude they are likely to invite from other category of farmers in a social system. Therefore it was felt necessary in initial debates and discussions by a team of scientists to include even agricultural labourers and rural artisans besides school going children in main stream of development.

Interventions

Project was implemented initially with five interventions namely:

1. Twenty two capsule of technologies which includes vermi compost, bee keeping, value addition and processing.
2. Providing timely information and training including linking of Village Resource Centres to Expert Centre established in collaboration with ISRO.
3. Ensuring required critical inputs for full scale adoption of improved technology including farm machinery on custom hire services.
4. Effective functional linkages to enlist synergy among related institutions.
5. Market intelligence to help farmers to derive best price for their produce.

Two pronged strategy namely; Integrated farming and addressing end to end issues were given due attention throughout implementation of project.

These interventions were implemented with active involvement of 32 scientists coordinated by a Project Coordinator and assisted by a Associate Project Coordinator. A team of eight young scientists (Research Associates) representing different subject matter areas were appointed to work with one thousand families each. Activities were implemented under close guidance and supervision of senior project team. Although four interventions were effectively implemented, intervention related to market intelligence was partially covered. Feedback analysis revealed that market intelligence was availed mainly by large farmers who generally produce large quantity at a time and it is economical for them to transport in a truckload at one go. In case of small farmers, salable farm produces are marginal in quantities, it was found uneconomical for them to hire a truck to transport, difficulty in transporting through public transport and addressing market dynamics individually is a challenge. Further, as majority of them are nuclear families, if husband goes for marketing, wife will have to look after entire family responsibility and vice-versa also. A strong debate and discussions were initiated for a period of six months with project staff, collaborating institutions and farmers to address this problem in beginning of second year of project. Thereafter, weighing all possible options, it was thought appropriate to promote Producers Associations to address marketing problem.

Start of Producer's Association

Efforts were made by project team with active involvement of farmers during middle of 2006 to start selected need based Producers Associations. This was an important milestone in project.

Ten associations were started for produces where problem of marketing was more and they are;

1. Rural Biofuel Growers Association
2. Jack Growers Association
3. Fruits & Vegetables Growers Association
4. Organic Farming Farmers Association
5. Flower Growers Association
6. Corn Growers Association
7. Federation of Women SHGs
8. Fish Farmers Association
9. Agro Processors Association
10. Chawki Rearing Centres

Although association was started with objective of enabling realisation of best price for produce, transport through organised group and bargaining for fair price, later it was realised that it could be productively used to address majority of backward and forward linkages. More specifically, promoting producers association was useful to manage critical inputs, technology, exchange of innovative experiences, farm machinery, credit, insurance, availing facilities from government, storage facility, transport, sharing labour, value addition and processing, joint study tour and so on to develop farm families' particularly small farmers on a sustainable basis.

Chawki Rearing Centre established by Youth

Low Vermicompost established by Youth

Apiary by a Small Farmer

Outcome of project

Project outcome reveals 11 percent growth rate in agriculture, more than two fold increase in net income for all farm families including small farmers with 2.52 lakh man days of additional employment generation annually. Agricultural labourers got increased employment opportunities besides engaging in value addition and processing, sheep rearing, poultry, catching fishes in tanks and marketing so on.

Farm Youth in Fish Harvesting

Dr. S. Ayyappan, DG, ICAR and Secretary, DARE visiting a small Farmer's feild adopted Integrated Farming

School Children involved in Biofuel momentum

Analysis revealed that outcome was due to timely empowerment and providing critical inputs free of cost, timely farm operations particularly for small farmers because farm machinery was provided on custom hire basis resulting in full scale adoption of improved technology besides additional support system and knowhow from all related institutions, which were effectively synergised. In fact benefit from these collaborating institutions was more than two folds compared to project cost. Number of enterprises for each families was increased generally from two to five which supplemented and complemented existing enterprises. More specifically apiary, vermicompost, diary, sheep, poultry, fishery, sericulture, baby corn, sweet corn, small scale floriculture, nutritional gardening, value addition and processing, direct marketing responsibility by small holders and so on have together contributed for added net income and additional employment opportunities particularly for small farmers.

Jack Fruit Nursery

Marketing of Jack Fruit through packing

Training of Farm Women on Preparation of Value Added Products

Sri H.D. Deve Gowda, Former Prime Minister and MP observing Value Added Products displayed and interacting with SHGS

More income and employment generation were realised due to start of 10 Producers Associations. Rural Biofuel Growers Association was started in project area with 42 rounds of meeting, visits and experience sharing which took nearly 18 months for commissioning processing plant. Farmers who were getting Rs 5 per kg during 2006 were paid Rs.15 per kg and procured through Milk Producer's Cooperative Society from respective service area with five percent overhead charges. Five unemployed semi skilled youth were given employment, payment was made timely and weighnent being transparent.

Many of family owning biofuel trees were hesitant to harvest produce before formation of Association, now every family members are actively involved in harvesting, old age people are engaged in dehusking and cleaning while school going children are involved in planting biofuel seedlings.

In case of Jackfruit, Jack Growers Association farmers were able to realise more than three fold added income for their produce. Many small farmers entered direct marketing of Jackfruit through Jackfruit melas in Lalbaugh and other places. These small farmers have realised four to five fold more income compared to bench mark year. Interestingly, many High School and College going students from these families have taken up direct marketing by selling bulbs since Jack harvesting season generally coincides with summer vacation.

Start of Fish Farmers Association was quite a good experience. Out of 321 water bodies, less than 20 were used for rearing fishes during pre project period. As a project intervention in first three years, fingerlings were supplied free of cost to all farmers owning individual water bodies including community tanks where group of youth were guided to manage. Further timely training inspired all of them to take up rearing fishes. Procurement of fingerlings fromHebbal/ Hessaraghatta was done through common transport arrangement and also distribution by project staff with help of volunteers from each panchayat. Marketing was done at Doddaballapur (near by town) on every Wednesday and Sunday in a specified place and time by association representatives. Payment to fish farmers was done in same day/next day in relation to quantity supplied and rate realised, which used to be nearly two folds compared to bench mark year. Consumption of fishes has increased several folds more so by small farmers.

Activity also provided additional employment to fish farmers and agricultural labourers. Fish Farmers Association was started during fourth year of project and thereafter, they managed enterprise independently.

Similarly, women federation for value addition and processing, marketing of fruits and Vegetables, sheep producers association and chalky rearing centre have made varying impact on income and employment generation. Interestingly all these associations are continuing in same wavelength despite project was withdrawn during March 2010. Follow-up work and needed guidance is provided by scientists of KVK in project area besides minimal follow up from university head quarters.

As a result of increased income among farm families, it was observed improved nutritional status particularly among children and women, increase in material possession such as mobile, motorcycles, household utensils and furnitures, construction of new house, renovation of existing house, behavioural changes such as communication skills, leadership qualities, change in social status suchas recognition through awards, using them as resource persons in KVK training programmes and other ongoing extension programmes. Some of office bearers and awardees were invited in adjoining districts, outside states, colleges and Universities to share their valuable experiences by paying respectable honorarium. These awardees were extremely happy about their recognitions and opportunities created expressing that they hardly dreamt. Overall standard of living of these families was improved significantly compared to baseline report fulfilling objective with which project was started.

Summary

Resources of small farmers are less compared to large farmers thereby they are unable to realise full production potential. Marketing is biggest challenge they find it difficult to address individually besides many social as well as psychological problems. In Indian social system, it is difficult to isolate small farmers and extend development support.

There are many ongoing programmes launched by both respective state and central governments exclusively meant for small farmers which could be dovetailed for their development. In project work, effective linkages were ensured by actively involving all development departments, financial, marketing and research organisations. Hence, there is need for addressing them together with other farmers for their overall development in Indian context.

In view of comprehensiveness of model, by promoting integrated farming, ensures independency for farmers to stand on their own and improving their confidence level, particularly through producers associations, more employment generation, emphasis on eco friendly technologies and above all respectable price for their produce, model has relevance. Therefore, there is absolute need for replicating this cost effective model with suitable refinement to suit different agro climatic regions and socio cultural mileau to bring about sustainable development of small farmers in India.

Reference

Reddy, A.A. (1971).*Extension Education*. Sree Lakshmi Press, Bapatla.

Singh D., and Singh, B.K. (2012). *Agriculture Extension and Rural Development*. Narendra Publishing House, New Delhi, India.

Gowda, N. K. (2011). *DBT-Rural Bio-Resource Complex Project: Final Project Report (2005-06 to 2009-10)*, University of Agricultural Sciences, GKVK, Bangalore.

Ray, G.L. (1991). *Extension Communication and Management*, Kalyani publishers, New Delhi.

□□□

Chapter-2

Relevance of small holder livestock production systems in India

K.T. SAMPATH

Former Director, ICAR-NIANP, Bengaluru

Small holder system covers more than two third of country's total production from livestock. In milk sector, although a good number of commercial dairy farms producing large quantities of milk in one place, have come into existence, intensive dairy units cannot replace small holder producers in near future and small units continue to play a significant role as milk marketing is well organized. The milk production may exceed requirement of domestic market in coming years and this calls for quality milk production for global trading. Major share of poultry meat and egg production in country is coming from commercial sector. However, this sector can also play a major catalytic role in promoting rural poultry production through contract farming. Encouraging small scale family based backyard poultry system by providing improved verities of dual purpose birds will help in poverty alleviation in rural sector. Unlike poultry sector, major share of sheep and goat meat comes from small holder producers. However, diminishing pasture land, unorganized market and exploitation by middlemen are some of areas of concern in this sector which requires attention of policy makers. Swine production is primarily a small scale unorganized rural activity which is manly concentrated in north eastern parts and tribal areas. There is a scope for converting this into a viable enterprise by providing desired support in improving productivity and marketing.

Livestock production plays a vital role in agricultural sector in India. Livestock sector contributes about 26% of agricultural GDP. Value output from livestock sector is about ₹ 733000 crores. Fast pace of urbanization, increase in income of people and awareness about protein needs have created a high demand for livestock products like milk, meat and eggs. Growth of agricultural sector during first four years of twelfth plan period is 1.6% against targeted 4%. However, livestock sector is growing at rate of 4.5%.

Small holder production system in India covers more than two third of country's total production from livestock. It covers landless, marginal and small farmers. However, growth of high value agriculture and structural changes in agricultural supply channels is sometimes causing hardship to small holder producers resulting in exclusion of some of them from system. This calls for a policy interventions to protect interest of small holder livestock farmers and to bring about entrepreneurship along with technological interventions for their livelihood security, nutritional security and overall improvement of rural economy. The present scenario and likely impact of intensive systems with regard to production of milk, meat and eggs are discussed in this paper.

Dairy Sector

India is largest producer of milk in world and produces about 162 million tones of milk annually. The major portion of this production comes from small and marginal farmers and landless poor, thus providing livelihood to about 70 million house hold. High intensive commercial dairy farms producing large quantities of milk in one place are less in number. The percapita availability of milk in country is about 300g per day which is as per recommended levels of Indian Council of Medical Research (ICMR). However, percapita consumption of milk in country is much below desired level compared to other developed countries, main reason being lack of affordability by some sections of population.

Fifty percent of milk produced in country is marketable surplus which is processed and sold as fluid milk and value added products.Remaining 50% is retained by Producers for household consumptions. Thus, small holder milk production system provides nutritional security to rural households in farm of milk for children, pregnant and lactating mothers in rural areas. Out of 50% of market surplus milk, only 30% is handled by organized sectors like milk cooperatives, producer companies and private companies. Remaining 70% is traded by unorganized sector. It is necessary to increase share of organized sector in interest of both producers and consumers.

Sale of milk is reasonably well organized. In the cooperative sector, milk societies at village level collect milk from the producers both in morning and evening. The milk thus collected is transported by milk unions to processing plants where it is pasteurized and sold to consumers either as liquid milk or as value added products like butter, ghee, cheese, ice cream etc. The payment is made to milk producers either weekly or fortnightly. Producer companies and private companies also follow more or less a similar system of milk procurement, processing and marketing. Now adays good number of dairy entrepreneurs has started large dairy farms with 100 to 1000 cows in different parts of country. Most of these dairy farms adapt latest technologies for feeding, management and milking of animals to produce quality milk and many a times get premium price for their produce. Organic milk is also produced by some of these entrepreneurs. Many times milk processors are at ease with these entrepreneurs in doing business. Thus, intensive system of dairy farming is making its presence felt. However, large scale dairy production system cannot replace small producers in near future. The small producers have advantage of competitive cost of production because of involvement of family labor and availability of by-products at house hold for feeding of dairy animals. Besides this, incentives by milk unions for their members in form of inputs like feed on credit basis, assured market for produce and uninterrupted income has added attraction for small farmers to stay in business. Thus small producers have a significant role in milk production in country. However, situation may not remain rosy for small holder producers in coming years. The percapita availability of milk in country has crossed ICMR recommended levels. Although a large number of value added products are being produced, production may exceed requirement of domestic market in coming years. Hence it becomes essential to explore export market. At present, less than 1% of milk/milk products are being exported. In view of stiff global competition from developed countries, it is essential that our producers are in tune with global standards with regard to quality of milk produced. Production of milk from disease free animals in hygienic environment becomes utmost important. Traceability is another factor which needs to be put in place to remain in global market.

Poultry Sector

Annually about 3.8 million tones of poultry meat is produced in country. The percapita consumption of poultry meat is about 2.8 kg per year against ICMR's recommendation of 11kg and global percapita consumption of 12kg per year. The annual egg production in country is about 75 billion. The average percapita egg consumption is about 70 eggs per year as against ICMR's recommendation of 280 eggs and global average of 155 eggs. Majority of this meat and eggs (75%) comes from organized sector.Poultry meat sector is growing at rate of 10 to 12% per year and layer production is growing at rate of 6 to 8% per year. The contribution of small scale producer is coming down year by year in poultry sector. The organized sector (commercial producers) has superior germ plasm, easy access to market and high quality management. Since returns are available in a shorter period of time, more and more commercial players are entering poultry meat sector and small producers are finding it difficult to compete with them. However, contract farming/integration system has helped some traditional poultry farmers to stay in business. Contract farming is a partnership between agribusiness/marketing firms and farmers. The firms provide day old chicks and feed at no cost to farmers and farmers receive fixed growing charges independent of market prices. The farmers are able to harvest six crops of broilers in a year. Thus, commercial sector with its technical and financial resources can play a major catalytic role in promoting rural poultry production as a practical and valuable option for poverty alleviation.

Targeting small scale family based backyard poultry system can be an area for poverty alleviation in rural sector. Non descript native fouls are mostly used in this system. Various types of dual purpose (meat and egg) germplasam with higher level of egg production and feed conversion ratio suitable for backyard poultry production have been developed by different research organizations all over country which can be used to boost rural poultry sector. Birds are allowed to scavenge partly and are supplemented with balanced diet. Marginal and big farmers can also be encouraged to start small scale intensive poultry farming in villages well connected by roads as a subsidiary occupation which helps in self employment to rural youth and women. High yielding commercial genotypes are recommended for this type of system. However, farmers need to be trained and provided technical knowhow by NGOs and govt. departments. These measures will play a big role in meeting targeted poultry meat and egg production in a shorter period of time.

Sheep and Goat Production

Unlike poultry meat, major share of sheep and goat meat comes from small, marginal and landless farmers. Sheep and goat contribute significantly to agrarian economy. Besides meat, sheep and goat also contribute milk, fiber and manure. They play a significant role in income generation, capital storage, employment generation and house hold nutrition. Common property resources (CPRs)–common grazing land, pasture and wasteland provide large proportion of nutritional requirement of sheep and goat in India. The resource free farmers depend on these CRPs as compared to rich farmers. However, in recent times semi intensive and intensive systems of sheep and goat rearing are being adapted because of decline in quantity and quality of CRPs. The performance under extensive system varies depending on quality and quantity of forage available. In some parts of country, sheep flock often migrate to nearby states as there is a shortage of grazing resources in their native tract. Under semi intensive system, sheep are allowed to graze in CRPs or cultivable /fallow land for 8 to 12 hours a day and then offered some concentrates. In this system, there are fewer fluctuations in production parameters than under extensive systems. The intensive system where sheep are stall fed with roughages and concentrates is immune to agro meteorological variations. Many entrepreneurs have also taken up sheep and goat meat production by rearing these animals under intensive system. However, small, marginal and landless farmers continue to play a significant role as these commercial ventures are not in a position to meet market demand of meat in near future.

There is very large potential in sheep and goat production for providing food security and economic development and gainful employment to resource poor people especially in arid and semi arid parts of country which need to be encouraged. However, scarcity of feed and fodder is becoming a major constraint due to diminishing common property resources. The village pachayats need to be given incentives to develop common grazing resources in villages, potential indigenous breeds need to be identified at micro level keeping in view local resource situations. The marketing system of sheep, goat and meat is not well organized. Small holder producers need to be directly linked with processors and consumers to help them from being exploited by middlemen. This calls for immediate attention of policy makers to organize marketing system similar to dairy sector by organizing sheep and goat rearer's societies. Technology adaptation is very low in sheep and goat production. The margin of profit can be enhanced by providing technological packages for scientific feeding and management of sheep and goat.

Swine Production

Pig production is primarily a small scale unorganized rural activity in India. It is an integral part of diversified agriculture particularly in tribal areas. In recent years a number of piggeries in private sector have also come up where pigs are reared on improved lines and manufacture of piggery products is carried out on scientific lines. The pig population in country is about 10 million, out of which one third is in north east region.

Pigs are reared mainly under three production systems such as scavenging system, backyard system and intensive system. Under scavenging system, pigs obtain their feed by roaming around and scavenging. The production performance is very low under this system. Under backyard system, pigs are housed in sheds made of locally available materials and are provided with kitchen waste mixed with vegetable wastes and other leafy materials.Intensive system of pig rearing is adapted by resource rich farmers, wherein crossbred or exotic breeds of pigs are reared on scientific lines.

Swine production is a profitable enterprise, which requires a comparatively small investments and income comes rather quickly.Pig production can be very useful in poverty alleviation especially in eastern states and tribal areas, given the desired support in improving productivity and marketing. Pigs are efficient converters of feed in to food for human consumption. Also there is a scope for pig production as a viable enterprise in villages well connected with metros where demand for pork is high.

Summary and Way Forward

The small holder producers play a significant role in milk production system and Intensive dairy units cannot replace them in near future. However, milk production may exceed requirement of domestic market in coming years. Hence it is essential to enhance global trade by producing quality milk. Major share of poultry meat and egg production in country is coming from commercial sector. Encouraging small scale family based backyard poultry system by providing improved verities of dual purpose birds will help in poverty alleviation in rural sector. The major share of sheep and goat meat comes from small holder producers and there is a wide gap between demand and supply. Diminishing common property resources, unorganized market and exploitation by middlemen are some of areas which need to be addressed to bridge gap. There is a scope for making pig production into a viable enterprise in north eastern region and tribal areas. This requires support in terms of improved germ plasma and scientific feeding and management. Overall, optimum utilization of livestock resources calls for transforming tiny and small holdings into production enterprise through infusion of technology and establishing direct market link between producers and consumers.

❑❑❑

Chapter-3

Small farmer production systems : Review of status and way forward

MRUTHYUNJAYA

Former National Director, NAIP (ICAR), New Delhi

Small farmers (owning less than 2 hectares) exist in huge numbers, and they cast their shadow over a whole range of development issues like food production, employment, poverty alleviation, etc. (Joachim von Braum, 2014; Shenggen Fan, et.al., 2015). There are nearly 500 million small farmers in world today and they are getting more numerous and smaller than ever before. Nearly 87% of world small farmers are located in Asia Pacific region. China and India together account for about two-third of small farmers, with those in India contributing about 50% of total agricultural output (Hazell, 2011). In India, during last three decades, number of small farmers increased by 77% from about 66 million in 1980-81 to 117 million in 2010-11. They now account for about 85% of all land holdings in India, with an average size of holding of 0.61 ha and cultivating about 45% (up from 26% in 1980-81) of total operated area. If current trend continues, it is estimated that India will have more than 155 million smallholders by 2021 controlling about 51% of operated area (Joshi, 2015).

The Unending Debate on Efficiency of Small or Big Farms

The relationship between farm size and productivity is intensively debated in India for many years. Earlier studies during 1960s and 1970s provide strong empirical evidences of inverse relationship; concluded small is beautiful and suggested land reforms, land ceiling and various other policies to support smallholders on ground of efficiency and growth (Ramesh Chand, 2011). But later studies during 1980s observed that inverse relationship is weakened and disappeared with application of modern technology, more use of critical cash inputs on large farms. Several recent studies, observed that on account of many developments like declining share of agriculture in GDP without accompanied by fall in share of dependency of workforce, rising disparity between per worker income in agriculture and non-agriculture and slowdown in growth of agriculture after mid1990s putting strain on income and livelihoods of small farmers, suggest policies to address twin problems of rising productivity and growth of agriculture as well as improving income and livelihood of small farmers. Ramesh Chand (2011) based on analysis of data during 1970-2006 observed, a) inverse relationship is more pronounced in recent times, b) efficiency gains with large farmers are not applicable to Asian countries, c) any move to increasing farm size will adversely affect productivity and growth of Indian agriculture and d) small farms are vulnerable and weak in terms of generating adequate income and sustaining livelihood. He therefore suggests reducing land inequality, lowering agricultural land ceiling and creating employment for smallholders outside agriculture but within countryside so that they get work and income from rural non-farm activities without quitting farms. But in a recent paper he observed that decent growth in farmer's income requires high growth in input, favourable prices for farmer produce, and some cultivators moving away from agriculture (Ramesh Chand, 2015). Shenggan Fan *et.al.*, stated that small is always not beautiful as small holders are not homogenous groups, they are diverse in size and other characteristics living in different ecologies and socio-economic settings (Pratap, et.al, 2017). As a result, he suggested that small farmers can prosper either through a move-up (through enhanced productivity, high value agriculture, and linkage with urban and global markets) or a move-out (safety nut activities along with upgradation of skills in short run and viable exit out of agriculture to seek non-farm employment opportunities in long run) strategy. Thus, messagefrom debate is that in case of small farmer development, no one size fits all and policies and strategies need to be diversity/situation specific.

Small Farms: Viability and Sustainability

Along with much debated efficiency of small farms over large farms, their viability and sustainability is also often debated (Joshi, 2015). They face several disadvantages like economies of scale, inadequate access to information, services and markets. Even their efficiency is argued to be diminishing with development of rising per capita income, migration of workers from agriculture, high production constraints making agriculture unattractive to young people depriving agriculture of energy, vitality and innovations, rising rural wages and cheap capital compared to land and labour and increasing climate and price risks. We often hear about small farmers trapped in a vicious cycle of food insecurity, malnutrition, poverty and some even reported to be ending with suicides. Though transition from small farms to large, commercial farms is a natural process of economic development, but it is not smooth and fast anywhere in world (Hazell, 2011). The cardinal principle of planned economic development should be that there should not be premature and rapid exit of small farmers who played dominant role in agricultural growth all along. They should be assisted to navigate smoothly and slowly. In this transition, role of Government, NGOs, private sector and every stakeholder is important to create a more equitable and enabling environment which includes assistance in targeted R&D and extension, small farm mechanization, forming effective marketing organizations, revamping of financial systems to meet small farmers credit needs, improved risk management policies, tenure security, efficient land markets, labour reforms, and if these fail targeted but time bound safety net programs (DAC, 2000). In addition, public sector should also adequately invest in provision of basic infrastructure, health, education and other human capital to improve market access and to include range of non-farm opportunities available to small farmers including planned permanent migration to urban areas. But cardinal principle in assistance, service and supplies is that they should be made self-dependent over a defined period of time after which there should be a clear exit /succession plan by outside agencies making local agencies/organizations owning/ mentoring them subsequently (Mruthyunjaya, 2017).

Renewed thrust on Smallholder farming with emphasis on improving income and livelihood

Lately developing countries, including India have been promoting smallholder farming owing to their critical importance in creating more employment, reducing poverty, improving food and nutritional security, improving income and livelihood security and their changing consumption patterns which help vitalize rural non-farm economy (DAC, 2007; Hazell, 2011). In view of economic development and labour becoming expensive, to harness advantages of smallholder farming, it becomes imperative to target needy farmers and area, introduce modern agricultural technology, institutional innovations, land reforms, promoting diversification, commercialization and value chain, linking farmers with markets and providing extension and credit and insurance support (DAC, 2000; DAC, 2007; Otsuka, et.al. 2014; Joshi, 2015; Ramesh Chand, et.al. 2015; Paroda, 2017; Birthal, 2017; Rao, et.al.2017; Narayanamoorthy, 2017; Mruthyunjaya, 2017; Ayyappan, 2017). When these are provided, examples abound that small farmers significantly improved their income and welfare and proved no more a liability on society/economy (ICAR, 2014 a; ICAR 2014 b; Ramesh Chand, 2017, Ayyappan, 2017). Ramesh Chand (2017) quotes a reference with several successful case studies of collective action by farmers working through organized institutions like Farmers' Producers Organizations. One distinguishing feature of successful examples/pilots is that there is dedicated, personal attention and passion for social service by catalytic agents/agencies/leaders of change. In this paper, an attempt is made to provide 2 sets of examples of success, one set steered by a public-sector research organization (ICAR) another by a private sector initiated but now an independent not-for-profit charity Syngenta Foundation for Sustainable Agriculture (SFSA).

Prosperity to small farmers in most backward regions of India under NAIP, ICAR

National Agricultural Innovation Project (NAIP) was implemented by ICAR with financial support of World Bank during 2006-2014 (ICAR, 2014 a; ICAR, 2014 b). The overall objective of NAIP was to contribute to sustainable transformation of Indian agriculture sector from orientation of primarily food self-sufficiency to one in which market orientation is equally important for poverty alleviation and income improvement through collaborative development and application of agricultural innovations between public research organizations, farmers, private sector and other stakeholders (ICAR, 2014, b). Out of 4 Components of NAIP, component 3 was to promote research on "Sustainable Rural Livelihood Security". In this component, a rural livelihood improvement approach is implemented in a consortium mode with 36 sub-projects covering small farmers located in 91 most backward districts distributed all over India. The project/consortium partners were 39 ICAR Institutes, 43 SAUs, 66 NGOs (4 even lead consortia) and 35 others including General Universities, Indian Institute of Technology, State and Central

Government Organizations and International Organizations located in India. The major technological backstopping was provided by scientists from ICAR institutes and SAUs while NGOs provided support in social mobilization, implementation of project activities, etc. Issues related to natural resource management, increasing farm productivity and profitability, constraints to sustainable and inclusive improvement of wellbeing of rural vulnerable groups (small farmers, women, SC&ST groups, etc.) and nutrition security, income improvements in a farming system perspective through a strategy of production interventions, value addition/processing, market/value chain linkage, and social interventions (non-farm activities, community resources, training, entrepreneurship, etc.) were addressed under 36 sub-projects of component. The special features of these innovations include: interventions appropriate to region, cluster approach in specific sites differing in natural resources, self-help groups formed for federating farmers, PPP to accelerate productivity and profitability, producer company through processing, value addition and marketing to realize higher benefits, rural technology centres (RTC) created for reducing knowledge gaps and technological empowerment, agri-business resource centres for entrepreneurship development, seed bank to make available quality seeds at right time, marketing linkages and value chain promotion for providing outlets for sale of produce and sustainability fund generated through farmers' contribution, development of inclusive community based organizations, linkages with Banks, insurance companies and other organizations, need based capacity development and creation of service providers, building chain and revolving fund, and formation of village level commodity/seed banks for post project sustainability. Major contributions covered:

a) Natural resource management.
b) Crop interventions.
c) Organic farming.
d) Horticultural crops.
e) Seed production.
f) Livestock, poultry and aquaculture interventions.
g) Farm mechanization.
h) Post harvest technology and value addition.
i) High pay-off interventions and
j) Sustainability of post-project activities.

A sample of successful production system livelihood achievement models developed under NAIP include following (ICAR, 2014 a)

1. Fish-pig-duck farming system in South Garo Hills, Meghalaya

Fish-pig (Hampshire)-duck (Sonali) based farming system was introduced in 11 villages of Sibbari cluster, South Garo Hills. Ten new ponds were constructed and 30 existing small ponds were renovated. Due to this intervention, fish productivity increased, improved ducks produced more eggs per year and number of furrowing and piglets per furrowing increased resulting in more income to farmers.

2. Integrated fish- makhana-water chestnut farming system in Dharbanga, Bihar

Multiple uses of water were attempted through integrating makhana with fish and water chestnut as a concurrent crop. The outcome of intervention from 96 beneficiaries tried in 50 ha area of Dharbanga, Sadar Block revealed that a total net income of ₹ 55,252 per year and employment of 341 mandays/ha/year as compared to only ₹ 18553 per ha and an employment of 223 mandays/ha/year with makhana as a single crop.

3. Integrated rice-fish-poultry farming system in Villupuram, Cuddalore and Nagapattinam, Tamil Nadu

Fish culture was taken up in trenches running along border of rice fields. Broiler birds, 1 bird/10m2 of rice area were housed in coop installed in fields so that chicken droppings fall into rice field. This enabled poultry waste to serve both as rice manure and fish feed. This resulted in addition of poultry manure -11.4 to 19.6 t/ha, suppression of pest-17-27% and increase in net return per household/annum of ₹ 6,600 to ₹ 10,100.

4. Land shaping in salt affected coastal areas of Sundarbans, West Bengal

Land shaping, construction of farm pond, ridge and furrow intervention. Earlier land was mono-cropped and only traditional rice varieties were grown in Kharif. After harvest of rice in December, land was kept fallow till transplanting of next paddy crop in June-

July. With intervention of farm pond technology farmers are practicing integrated farming system (crop-vegetable-fish) with higher farm income.

5. Redgram transplanting/dibbling in Bidar, Karnataka

Improved variety of redgram is cultivated by transplanting/ dibbling method. The advantages of redgram transplanting are advance in sowing, less pod-borer attack, drought resistant, 2-3 fold increased branching and increased yield with a net profit of was ₹ 66, 125.

6. Precision farming of marigold in Tamil Nadu and Karnataka

Adoption of precision farming method developed for African marigold hybrid L3 resulted in yield increase as well as in xanthophyll content leading to farmers realizing higher price and farm income.

7. Tamarind processing in Bastar, Chattisgarh

In Pedawada cluster of Bastar district, 60 tribal families were organized in 6 groups for collection, dehulling, deseeding, packaging and marketing thus shelving middlemen. Aproduce of about 200 quintals was handled by cluster which provided additional income of about ₹ 8,200 to each family annually. Community organized group procurement, processing and marketing increased profit up to 40-50%.

8. Value addition of non-timber forest produce in Godda, Jharkhand

i) **Leaf plate mechanization:** Optimum utilization of forest produce through value addition by promoting leaf plate machines (5 SHGs-15 persons each SHG) generated a net income of ₹ 0.17 lakh/machine in three months.

ii) **Pickle making:** Group were trained for value addition (pickle making) of abundant underutilized jackfruit of Sundarpahari and Godda forest ranges and sold it under brand name of "Yogini", produced 2 tonnes, fetched higher prices, employment to 117 women and additional income of about ₹ 1150 to every participant annually.

9. Sharing of groundwater through pipeline networking and social mobilization in Rangareddy, Andhra Pradesh

Seven borewells owned by different farmers located over a 45-acre patch were linked through a network of underground pipelines. The groundwater extracted through these borewells was shared by 18 farmers through sprinkler irrigation. This enabled farmers to diversify their cropping pattern during kharif and able to provide protective irrigation and save crop during mid-season droughts as well cultivate crops during rabi season preventing outmigration of farmers.

10. Seed Societies in districts Jhabua and Dhar districts, Madhya Pradesh

Two Seed societies with 22 and 21 farmers as members were formed (Laxmi Beej Utpadak Sahakari Sanstha Maryadit, Golabadi and Sharada Beej Utpadak Sahakari Sanstha Maryadit, Narasingrunda, Rolta) to accelerate availability of quality seed in this tribal district. They did well and presently there are 21 seed societies are working in Dhar District alone offering services to tribal farmers to increase their income.

11. Collection and marketing of farm produce in Kerala

Vegetables and Fruits Promotion Council (VFPCK) started fruit and vegetable collection points in Waynad district of Kerala. This is managed by 17 farmers on self-help mode. 17 farmer owned procurement and 2 retail centres established with handling capacity of about 2500 tonnes. They provide transportation facility to distant markets. It has taken GI registration for speciality rice varieties Jeerakasala and Gandhakasal.

12. Mushroom cultivation in Odisha

Mushroom cultivation was successfully introduced using agricultural waste in Dhenkanal, Khandamal and Kalahandi districts of Odisha. This resulted in monetary benefits to about 104 small and marginal farm families.

In all innovations under 36 sub-projects of Component 3 of NAIP, 2 lakh farmers directly and an equal number indirectly through horizontal spread (demonstration effect) through technology, institutional and social engineering innovations were guided to increased income, better quality of life, livelihood and nutritional security. At overall project component level (Component 3), it yielded financial benefit cost ratio (FBCR) of 1.91 and economic benefit cost ratio (EBCR) of 1.67 indicating economic and financial feasibility of investment in these models.

Prosperity to Smallholders through Commercial Agriculture promoted by SFSA, India

Syngenta Foundation for Sustainable Agriculture (SFSA) is not-for-profit Swiss Charity based in Basil, Switzerland. The mission of Foundation is to find ways of increasing income opportunities of small-scale farmers and it wanted to play its role in India as majority of Indian farmers are small (Partha R. Das Gupta, 2017). In 2004, a core team with Dr. Partha R Das Gupta, former Country Director, India based Syngenta Company as a member was formed to draw up a plan. The Foundation was established in 2005 as a separate legal entity to take care of other project activities. The Foundation began with hypothesis that income of smallholder farmers can be raised by first providing them with knowledge of advanced technologies that can help raise farm productivity; this could be followed up by improving their access to markets so that they can earn bigger share in value chain. To validate this hypothesis, Foundation undertook projects in four carefully chosen areas of country which were economically disadvantaged and backward as projects under NAIP were selected. The following is account of achievements of Foundation in first 10 years of establishment narrated by Dr.Das Gupta who was personally associated and mentored projects right through. As we see through these pilot projects implemented one after earlier by learning and growing in different parts/pockets of India, they prove hypotheses set at beginning that it is possible to increase income of small farmers through knowledge, technology and access to markets.

The following case studies provide some additional hopes, models, principles, opportunities for smallholders' development in diverse, harsh ecologies and poor socio-economic settings in most disadvantaged regions of India.

1. The Anandwan Story-How Agriculture Evolved as its Major Economic Strength

Anandwan is a village near Warora, a tahasil of Chandrapur District of Maharastra in central India. The local partner was Maharogi Sewa Samithi (MSS) managed by Baba Amte. The Foundation planned its intervention on agricultural farm through use of improved technology. It also planned interventions on a larger farm of MSS at Somnath around 100 km from Anandwan. The pilot program was demonstrating high value vegetables using hybrid varieties. The pilot was a great success and it went into a project mode with supply of some seeds, drip lines in patches and technical guidance. The farmers were impressed with success in pilots and planned more demonstrations. The Foundation wanted to make Somnath a profitable commercial farm. It augmented irrigation facility by building a reservoir to provide protective irrigation. The farm produced higher yields of rice in kharif and much higher yields of vegetables and other field crops in rabi. Inter-connecting of water bodies was done and crops were taken in summer season also. SRI system of rice cultivation was followed which was also a success. Gains in growing vegetables and field crops continued during second year. Diversification into seed multiplication and other crops taken up in third year. From successful farming, strengthening extension services was pursued to share success with neighbouring farmers in nearby villages and many adopted innovations and got benefitted also. During next phase, Foundation assisted MSS in starting vocational education school and follow also new approach to extension, namely market-led extension. Foundation helped MSS also to start "Anandwan Institute for Transfer of Agricultural Technology" which started working from 2010 and it became Anadwan Agri-Polytechnic in 2013. In 2 years, Agri-Polytechnic became self -sustaining, premier institution and Somnath turned into a large and profitable commercial, professionally managed agriculture farm in central India. The Foundation exited in 2014 but MSS continuously made some additional developments like substantial increase in irrigation resources, levelling and consolidation of fields and farm operations were professionally managed which was initial vision of Foundation.

2. Resurgence of Agriculture in Jawhar- Bitter Gourds Bring Back Farmers to their Sweet Homes

Jawhar is a tribal inhabited Tahasil of Thane district in Maharashtra state. Though it receives heavy rains during monsoon season, it becomes dry during summer. Foundation has constructed a check dam and it wanted to explore possibility of utilizing its water for irrigation of nearby farms. The farmers were following primitive agriculture with low yields by growing rice (main crop), ragi, jowar and niger. In some pockets with availability of water, they grew vegetables in dry rabi season. The soils in upper and mid-valleys were highly degraded by runoff and erosion. They grew crops in kharif and left land fallow afterwards. Agriculture just provided them some grains but to secure livelihood, they migrated to nearby areas with their families and worked as unskilled labours. It was learnt that their main constraints were less water, no cash to buy agricultural inputs and limited access to knowledge. Farmers were very reluctant to grow vegetables as it was costly and more-risky. It was planned to disseminate knowledge of appropriate technology, demonstrate benefits of irrigation, introduce HYV of field and vegetable crops. Pilot project working with Pragathi Pratisthan, locally popular NGO began in 2006-07 with farmers' training meeting followed with distribution of certified seeds, demonstration of GAPs, trials in farmers' fields of new crop varieties and use of special agro-techniques like SRI. Hybrid rice seeds were also used. The response was good and farmers quickly learnt from fellow progressive farmers also and realized higher yields. A kiosk (ATIC) was opened for answering farmers' questions on use of modern technology. To provide quality seedlings to vegetable growers, a few low-cost poly houses were built by project. New alliances with BAIF as a partner and extending coverage to nearby villages, during 2007-8, farmers were assisted to improve yields of rice and vegetables and they were really inspired by success. In next phase after rising productivity in first phase, collective marketing was tried by tying with BAIF and Amrai Tribal MITTRA Fruit Processing and Marketing Cooperative Society. Farmers realized better prices and thus encouraged. Next year they went for independent initiative to tie up with other farmers' groups in nearby areas also to take up collective production and marketing of vegetables. Horizontal expansion to neighbouring Tahasils and diversification and reforms in working with Producers' Associations was promoted during 2009-10 and with that success increase in area under vegetable cultivation particularly bitter gourds whose demand was increasing took place. Tying with a private company engaged with exporting vegetables was also attempted along with training farmers in Global GAP norms. Thus by 2014-15, by collective growing on a commercial scale and marketing of vegetables, Jawhar (Konkan region of Maharashtra) became a major vegetable growing belt/vegetable valley involving more than 10,000 farmers catering to nearby Mumbai and other wholesale markets. Another major development in this transformation was substantial increase in number of input dealers operating in project area contributing to overall agri-business development of region.

3. Emergence of New Kalahandi-Its Journey from Basket Case to a National Seeds Hub

Kalahandi along with its contiguous districts of Bolangir and Koraput in Orissa state was considered as a major food insecure region of India. Foundation in partnership with Karrtabya, a local social service and development activities NGO decided to work together during 2006-7 to benefit tribal farmers in Kalahandi with focus on transfer of technology and crop diversification to raise farm productivity and farmers' income. The area had poor soil, degraded as a result of surface run-off during rains with erratic monsoons as well as due to mining for semi-precious stones resulting in poor rice crops leading to food shortages. The situation started improving with Indravati irrigation project completed in 2001. They could now take rabi crop but yields remained low. Farmers were not following modern agricultural practices and growing few vegetables with outdated varieties. Their main concern in vegetable production was lack of market demand. The Foundation along with Karrbatya organized a workshop by inviting expert scientists from research organizations and officials of development departments and provided training on modern agricultural practices, importance of quality seeds, management of pests and diseases, availability of inputs, etc. The workshop was a first move towards technology driven agriculture. Low cost poly houses were erected to assist cultivation of high value hybrid vegetables by supplying quality seedlings. The pilot during 2006-7 was highly successful and new directions and bigger goals were set for next year to raise productivity of rice, special emphasis on growing high value vegetables, increasing project outreach and collaborate with public sector to increase impact of project. SRI cultivation was promoted in 2007-8, public-private partnership was promoted, exposure visit of project farmers were organized and success in second year was again highly impressive. In 2008-09, clusters of vegetable villages extending coverage of area to 12 blocks of Kalahandi district were formed and linkage with market/supply chain was promoted. The success in third year were of greater magnitude. In fourth year (2009-10), goal was to make project/programs sustainable by, i) integrating vegetable growing with commercial hubs to be managed by producers themselves on a profit sharing basis and ii) engaging farmers in production of high value seeds. Over all, by 2013-14, seed production (vegetables, hybrid rice and maize) was well taken and farmers' hubs were also successful as anticipated and thus Kalahandi has emerged as a viable seed production centre of country increasing prosperity of Kalahandi farmers.

4. Small Farmers Make It Big in Bankura

Bankura is one of distressed districts of West Bengal where average size of holding was smallest and highly fragmented, crops occasionally suffered losses from drought. Though it was receiving good rainfall, it was lost as run-off over an uneven land surface, poor water-holding capacity of soil and lack of proper soil and water conservation measures followed in area. The Foundation joined hands with NGO Shamayit Math which had done some good social service work to improve quality of life of rural poor. The goal of pilot was to provide advanced knowledge to enhance productivity and eventually increase income of farmers. Kharif rice was main crop and because of water availability, there was no crops after kharif. A workshop was organized for farmers during June,2006 by involving scientists and extension staff of departments and training was given on new technologies, modern agronomic practices to increase productivity of rice and cash crops (potato, tomato, etc.). Farmers took up demonstrations and fully convinced about benefits. Many farmers participated and results of pilots of first year were very impressive. The project was fully launched in 2007-8 to enhance productivity of rice, scale up seed multiplication through farmers' participation and augmenting farm income from other sources like goat rearing, duckery and fish culture. The results of 2007-8 were encouraging but suggested scaling up to enhance overall impact. During 2008-9 and afterwards, diversification of project dimensions were pursued by emphasizing market-led extension with a goal of establishing collective production and marketing of vegetables based on hub-and-spoke model, bringing high value hybrid vegetable seed production operations (brand name of "Sree Rohi" seeds), alliance with Micro-Finance Institution, BASIX, linking potato growers with processing industry, commercial growing of vegetables with credit tie up with a Bank. By 2014-15, program was running on its own with minimal back up support from project and thus project could put Bankura on India's map of hybrid vegetable seed production centres.

The overarching factors which triggered transformation inlivelihood of smallholders in these 4 case studies are:

a) Knowledgeand use of modern technologies,
b) Improved access to markets/value chain
c) Market-led extension
d) Personal involvement of a professional with passion for social service since beginning of project as a change agent and mentoring along with others and continued interest in their sustainability beyond formal exit of projects from project area.

Way Forward

The most distinguishing common features of NAIP sub-projects narrated earlier as well as 4 case studies of SFSA and many successful cases referred to in some citations of this paper from point of view of way forward for agencies particularly interested in smallholder development in India like Small Farmers Agri-Business Consortium, NABARD, etc. in my view are summarized below:

1. Strong and sustained commitment and passion for social service by change agents/project authorities as catalysts of change in process of development even at personal and institutional inconveniences and an equally intense trust and cooperation, faithful practice/follow-up by farmers and community
2. Proper selection of target area and farmers and clear understanding of socio-economic, agro-biological settings and local/rural and dynamics of target farmers, community, stakeholders, region
3. Review of past efforts and results to change at each location, assessment of potentials and options still available but not utilized; key institution/agency involved/can be involved and priorities for change to prepare tailor made plan of action for specific situations/sites through wide consultation with all stakeholders

4. Focus and clear goals; selecting a reputed, credible local partner well connected with grassroots, particularly small farmers who can serve as an incubator and later become a role model/potential successor to subsequent time bound project exit; changes through trial and error but persistent efforts; step by step progress with incremental changes; learning and growing; pilot projects with interventions which are replicable and sustainable and learning from progressive neighbours; look for new alliances and diversified innovations with coordination, convergence of competencies and clear understanding of sharing resources and credit; scaling up to establish proof of concept; close monitoring, on-course correction including trimming work portfolio (if needed) outside goal/scope by project team; involving extension and other relevant line departments including private companies in planning and monitoring; proper use of star/champion farmers for spread of successful innovations; clear plan of exit/phase out plan and a careful succession plan; persistent efforts by change team and individual and collective action by farmers to convert adversaries to advantages.
5. The project initiatives not only improve wellbeing of target population but also will have multiplier effects to trigger agri-business (eg. Input industry, rural banking, insurance, rural services, etc.), development ecosystem of region.
6. Comprehensive and sustainable development takes place only when community itself gets actively involved, change mindset and practice agriculture as an enterprise and owns up and constantly learns from successes and failures.

Overall, We have enough success stories scattered all over India attempted by many agencies and individuals for improvement of smallholder agriculture neatly conforming to theory and sources of growth, namely, a) improvement in productivity, b) enhanced resource use efficiency, c) sustainable increase in enterprise intensity, d) diversification towards high value enterprises, e) planned shifting/move-out towards rural non-farm employment activities with needed skill upgradation, f) initiative to explore institutional innovations and access all available development opportunities (knowledge, technology, skill, inputs, credit, insurance, market, development/safety net programs of Government and other agencies, etc.). But in a large country like ours, number/scale of present attempts towards smallholders' development are less/small and needs to be expanded/scaled up through concerted action with maximum coordination, convergence, commitment and passion to realize goal of improving smallholders' income and welfare at a faster pace.

Reference

Ayyappan, S. (2014). Family Farms: Farm, Feed and Flourish, Presidential Address, Foundation Day and 21st GB Meeting, 5th June 2014, National Academy of Agricultural Sciences, New Delhi.

Ayyappan, S., Letha Devi G, Subhash, S, Devi, MCA, P.K.Dixit (Eds.)(2016). Proceedings of National Consultation on Small Farmers' Production Systems: Way Forward, 22-23 December, 2016, ICAR-NDRI, SRS, Bangalore., pp. 74.

Braum, J.V. (2014). Small Farms: Large Numbers, Great Diversity, Big Role for Economic Transformation and Innovation. Keynote Paper, 8th International Conference of Asian Society of Agricultural Economics, Dhaka, October 15-17, 2014.

Birthal, P.S., Negi, D.S., and Roy, D.(2017). Enhancing Farmers' Incomes: Who to Target and How? *Policy Paper 30*, ICAR-NIAP, New Delhi.

Chand, R., LaxmiPrasanna, P.A. and Singh, A.(2011). Farm Size and Productivity: Understanding Strengths of Smallholders and Improving their Livelihoods.*Economic and Political Weekly*. June 2011. XLVI (26&27): 5-11.

Chand, R., Saxena, R., and Rana, S.(2015). Estimates and Analysis of Farm Income in India, 1983-84 to 2011-2012.*Economic and Political Weekly*. May 30, 2015. L (22): 139-145.

Chand, R. (2017). Doubling Farmers' Income: Strategy and Prospects. Presidential Address, *Indian Journal of Agricultural Economics*. 72(1): 1-23.

DAC. (2000). National Agricultural Policy, Department of Agriculture and Cooperation, Ministry of Agriculture, GoI.

DAC. (2007). National Policy for Farmers, Department of Agriculture and Cooperation, Ministry of Agriculture, GoI.

DasGupta, P.R. (2017). Commercial Agriculture by Indian Smallholders, Maya Publishers, New Delhi.

Hazell, P.(2011). Five Big Questions about Five Hundred Million Small Farms. IFAD Conference on New Directions for Smallholder Agriculture, 24-25 Jan, 2011. IFAD,Rome.

ICAR.(2014a). Strengthening Agricultural Development and Livelihoods, Project Implementation Unit, NAIP, ICAR, New Delhi, May 2014.

ICAR.(2014b). An Initiative Towards Innovative Agriculture, Final Report-National Agricultural Innovation Project, Project Implementation Unit, NAIP, ICAR, New Delhi, December, 2014.

Joshi, P.K., (2015). Has Indian Agriculture Become Overcrowded and Risky? Status, Implications and Way Forward, Presidential Address.*Indian Journal of Agricultural Economics*. 70(1): 1-41.

Mruthyunjaya.(2017). Doubling Farmers' Income by 2022: A Reality Check. Keynote Address, Biennial Conference, Indian Society of Weed Science, 1-3 March 2017, MPUAT, Udaipur.

Narayanamoorthy, A. (2017). Farm Income in India: Myth and Realities, Conference Keynote Paper.*Indian Journal of Agricultural Economics*. 70(1): 24-48.

Otsuka, K., Liu, Y.Y., and Yamauchi, F. (2014). Future of Small Farms in Asia, 8th International Conference of Asian Society of Agricultural Economics, October 15-17, 2014, Dhaka.

Paroda, R.S. (2017). Re-orienting Agricultural Research for Development Agenda for Sustainable Livelihood Security, Strategy Paper, Trust for Advancement of Agricultural Sciences, TAAS, Pusa Campus, New Delhi.

Rao, N.C., Sutradhar,K., and Reardon, T. (2017). Disruptive Innovations in Food Value Chains and Small Farmers in India. *Indian Journal of Agricultural Economics*. 72(1): 24-48.

Shenggen,F., Brzeska, J., and Olofinbioyi, T. (2015). Helping Small Family Farms to Move-Up or Move-Out. *IFPRI Food Policy Report-2015*. IFPRI, Washington, D.C. USA.

□□□

Chapter-4

Smallholder dairy farming system: Issues and way forward

SATISH KULKARNI[1] AND MUKUND A. KATAKTALWARE[2]

[1] Former Head, ICAR-NDRI, SRS, Bengaluru

[2] Senior Scientist, ICAR-NDRI, SRS, Bengaluru

India is the largest milk producer in the world producing about 155.5 MT (2015-16) of milk annually accounting for 18% of world milk production. The milk in India is produced by a large number of farmers who are dependent on dairying for their livelihood. It is estimated that nearly 68 million households are involved in milk production along with other agricultural and animal husbandry practices or other non-agricultural occupation. The contribution of millions of small farmers in milk production in India is a great success story worth emulation by other countries where agriculture provides a major livelihood option. Since dairying is relatively well organised than other agricultural produces, there is scope and opportunity for further growth in this sector. Also combining dairying with other agricultural and animal husbandry practices, as part of mixed farming is likely to provide better financial security to the farmer. The rural women handle a large part of the dairy operations, for whom, it is a tool for financial empowerment. In recent time, it is observed that dairy development is progressing well in States, which were lagging in milk production especially in the eastern states like Jharkhand and Bihar and there is interest in the north-eastern regions.

Dairy Farming Scenario

The small and marginal farmers contribute about 70% of total milk production along with other agricultural produces contributing for the diversity of food production. A network of about 198 dairy cooperative milk unions have covered about 15.83 million farmers under the ambit of 1,70,992 village level dairy corporative societies.

About half of the milk produced is retained for household consumption and only 30 to 35 percent is delivered to formal milk processors. Inspite of unprecedented drought experienced by most of the States, milk production is growing at about 6.5% annually for the preceding two years as against the average of 4.7% over the earlier decade. A study on profitability and efficiency of dairy farms in India conducted by Delgado *et al.* (2003) indicated that smallholders have higher profits per litre of milk than large-scale producers still they could be driven out of the market because large farmers are more efficient than smallholders (regional variations: small more efficient in western region and large in northern region), large farmers produce larger volumes, smallholders have difficulty complying with quality standards and smallholders have less access to world dairy markets. As observed in the above-mentioned study, in the northern region, large farmers receive higher prices than small farmers; while in the western region, average prices received by small and large farms are almost the same which indicates that cooperatives benefit small-holders more than large producers. As food crop production has almost reached a plateau in terms of production and productivity, declining feed and fodder resources, constantly increasing demand for milk and milk products, there is an imperative to take up dairy farming as a viable commercial venture to meet the domestic requirement and to exploit the export potential of indigenous milk products. In Punjab and Haryana, it has been well adapted by a number of farm families as a full-fledged business. However, there are several constraints, which could play spoilsport in sustainable growth of the sector. The major constraints faced by the small and marginal dairy farmers in all milk production systems are related to the productivity of animals, cost of feed in relation to the procurement price, availability of grazing land or green fodder, lack of mechanisation in milking, veterinary care and appropriate breeds depending on the milk production system and the area of operation.

In order to keep the pace of growth in milk production the following issues to be flagged as way forward:

i) Supply of ingredients at a competitive price to feed manufacturers on the pattern of allotment of wheat to bakery industry.

ii) Development of appropriate breed of milch cattle, which suits the different regions and different category of farmers.

iii) Developing appropriate community-based grazing land for the benefit of dairy farmers.

iv) Assured purchase of milk throughout the year.

v) Skill development and cold chain infrastructure.

The time has come when the farmer and his next generation needs to be incentivised for continuing the farming operation by extending special and innovative schemes. Some of the possible approaches in this direction are detailed below.

Feed and Fodder issues

Feeding of Dairy Cattle forms the major expenditure for the dairy farmer. It is estimated that nearly 58-65% of the value of milk is spent for feeding the cattle. In this direction, many steps have been initiated to supply balanced feed ration, specialised feeds to meet the micronutrient deficiency and other varieties of feed to the farmers through the milk Cooperatives. Even large dairy plants in private sector have provisions to supply compounded feed at a competitive price to farmers supplying milk on regular basis.

Indian Cattle Feed Industry is also growing at around 7% keeping pace with the growth in Milk Production. The need of the day is to reduce the cost of these feeds so that the Country continues to be the largest producer of Milk at most competitive price.

In this connection the way forward could be:

a) **Supply of Grains at a competitive price:** India is producing sufficient cereals to meet the demand from its population. The Government also procures cereals from the farmers by paying Minimum Support Price and stores in its warehouses to meet the demand from the below poverty line consumers and also to maintain the emergency requirement of 30 million tonnes. However, many times, due to poor storage facilities the deterioration of the stored grains starts and at that stage, the poor quality of these stored grains is diverted to the cattle feed industry. This reflects certain drawbacks in the estimation of the demand and this needs to be corrected so that from the beginning itself a portion of the cereals is diverted towards the feed industry to meet the requirements of the Livestock which is main livelihood support for the farmers in the country. Currently, the wheat is allotted to bakery industry at acompetitive price to maintain the price line of bread and other staple food. This principle needs to be applied even to the cattle feed manufacture.

In this way, it should be possible to maintain the cost of feed supplied to the farmers through the cooperatives.

b) **Encouraging fodder production**: It is observed that for the past many years India has a good stock of cereals to meet the demand of its population and the milk production and consumption is increasing. Now the time is ripe to encourage farmers to reserve a part of their land for growing fodders which is in short supply. Today even the fodder is fetching good returns to the farmer and this needs to be further intensified by the Govt. In the case of losses to the farmer due to the growing of fodders, then the Government needs to come out with certain schemes to compensate the losses to encourage the farmers. Also, farmers need to be skilled for value addition through Silage preparation, bailing or complete feed block production so that they will be able to recover better returns on the green fodder. If and only if the farmers get remunerative returns from their land holdings they will continue with the practice of cultivating and growing fodders. This also is less risky to the farmer as in the event of failures of monsoon the losses occurring will be lesser in comparison the growing of cereals.

c) **Feeding of Calves**: Optimum growth in the body weight of calves is essential for the future dairy development. This requires proper feeding after weaning the calf. However, not many of the farmers are aware of the importance of feeding the calves for attaining optimum body weight. In recent times many of the cooperatives and private feed manufacturers are providing calf starters and these are being supplied at a competitive price to the farmers. This type of schemes needs to be encouraged by the Govt through Incentive schemes.

Development of Appropriate Breed of Milch Cattle

India is world's largest milk producer, however; productivity per animal is very low. The average yield (kg/day) in exotic cows, crossbred cows, indigenous cows, non-descript cows, indigenous buffalo and non-descript buffalo in 2015-16 was 11.21, 7.33, 3.41, 2.16, 5.76 and 3.80 respectively (Annual Report, DAHDF, 2017). The low productivity of bovines in India is due to the gradual genetic decline and general neglect of animals, shortage of feed and fodder, and poor reproductive efficiency of dairy animals (Kataktalware *et al.*, 2017).

Keeping this in view, the National Dairy Plan I, a central sector scheme of Government of India launched in 2011-12 being implemented by National Dairy Development Board and a new scheme namely "National Mission on Bovine Productivity" launched by the Govt of India in 2016 are aimed to increase productivity of milch animals and thereby increase milk production to meet the rapidly growing demand for milk. The National Mission on Bovine Productivity will be implemented as a part of Rashtriya Gokul Mission under Umbrella Scheme White Revolution-Rashtriya Pashudhan Vikas Yojna.

Providing quality AI services at reasonable cost in real time at every farmer's doorstep is the foremost challenge for rapid genetic improvement of animals. Further, creating awareness among farmers and AI personnel for usage of suitable breed semen in view of resources of the farmer, agro-climatic conditions and breeding policy of the State is an issue which needs to be addressed to avoid higher level of exotic inheritance beyond permissible limit and to ensure purebred indigenous germplasm is maintained. Dairy development is likely to face many challenges due to climate variability and the requirement of appropriate breed of milch cattle which can produce milk under adverse conditions of feed & fodder availability, thermal discomforts due to variation in temperature conditions and with good disease resistance has increased in recent times. In this direction, it is needless to mention that the country has been endowed with reasonably good Milch and dual purpose breeds of Cattle like Gir, Kankrej, Sahiwal, Hariana, Rathi, Tharparkar, Ganga Teri and buffalo breeds like Murrah, Surti, Mehsana, Jaffarbadi etc. These breeds have good ability to produce a reasonably good quantity of Milk under adverse conditions and the mean production in a lactation period is about 2800 - 3000 litres which is nearly 2.5 times greater than the current mean of milk production per cattle which is around 1500 lits per lactation. Hence it is the timethose good indigenous cattle needs to be used for selective breeding and for up gradation of native stock with superior indigenous cattle. Further, for progressive farmers apart from high yielding indigenous cattle, appropriate breeds needs to be developed by upgrading the best of the local breed by using semen from the best of indigenous cattle breeds like Rathi, Gir, Sahiwal or Kankrej. Such breeds developed should be able to produce at least 5000-6000 lit of milk per lactation.

India has one of the largest breeding infrastructures in the world with 57 Semen Stations and more than 3000 breeding bulls. Semen

Production has increased from 22 million straws (1999-2000) to 97 million straws (2015-2016) and the number of inseminations has increased from 20 million to 69.29 million (Annual Report, DAHDF, 2017). It is this facility, which needs to be utilised for providing high milk yielding indigenous breeds that should catalyse the milk production in the country in the coming years.

Milk Processing and Marketing

In India currently, about 43% of the milk produced is utilised at the household level and about 28% is processed through organised dairy sector leaving about 29% of the milk which is processed through the Traditional Milk Marketing System. In this 29%, about 50% is accommodated by the urban producers who are not going to part with their milk for organised sector procurement. It is high time that through appropriate legislation this sector needs to be controlled as in the urban areas the cattle will be allowed to graze and this creates a lot of bottlenecks on the streets of urban areas which have already been chocked. Further, these cattle which are of good genetic material consuming plastic and other toxic wastes from the garbage posing possible health hazard situation. This menace can be controlled only through the 'will' of the Govt. by appropriate legislation.

The milk is easily available for procurement and processing is about 15-16% of the total milk production, which available in scattered geographical regions with poor infrastructure and in areas where the dairying is not well developed. In this direction, the Eastern part and North East parts have good scope for organised dairy development. If India has to achieve at least 5% growth, it is these states should be brought into the fold of dairy development.

The marketing of milk and milk products have seen a sea change in recent years with nearly 20 variants of milk packed in more than 40 stock keeping units is a great development which has occurred in the country. Additionally, there are more than 40 different dairy and milk-based products which are being marketed in the country and this market should expand to underserved areas of the North East India, Andaman and Nicobar Islands. The value addition is of paramount importance if the dairy farmers are to be adequately compensated for their production and the Govt. or authorities should encourage the products and value-added products with newer schemes for dairy development.

The various Govt schemes operational currently are:

- National Programme for Bovine Breeding and Dairy Development
- National Dairy Plan
- National Livestock Mission
- National Mission on Bovine Productivity
- Rashtriya Krishi Vikas Yojana
- National Rural Livelihood Mission
- Schemes for Cold Chain, Value addition and preservation infrastructure (MoFPS)
- National Mission on Food Processing
- Support to training and empowerment programmes for Women (STEP)
- Grant in Aid
- Special Central Assistance to tribal Sub-Plan and scheme of Grant in Aid to State Govts. (ministry of Tribal Affairs)
- NABARD Warehousing Scheme
- Financing and Supporting Producing Organisations (NABARD)

Apart from these, there are considerable numbers of State schemes to support the dairying as a vocation.

Skill Development

In order to achieve the growth of dairy development at envisaged rate of 5% it is necessary to develop the requisite skill. The areas where the skill development required are:

- Cattle management
- Veterinary health
- Milk processing
- Product manufacture
- Dairy extension
- Dairy engineering

- Quality assurance
- Economics and management
- Genetics and breeding

The skill development is required at all levels from the three months certificate programmes to Doctoral levels. In this direction, the Govt. needs to invest so that there is no shortage of required skills at the time of programme implementation.

Summary

It is summarised that there is vast scope for developing dairying in the country, which may serve for the socio-economic up-liftment and livelihood of dairy farmers. The areas that need to be focussed are appropriate breed development, feeding management, milk procurement & marketing and lastly the skill development. It is possible to grow in milk production at 5 % for the next decade if we can address these areas of dairy development.

Reference

Annual Report, 2016-17. Department of Animal Husbandry, Dairying & Fisheries Ministry of Agriculture & Farmers Welfare.

Government of IndiaDelgado, C., Narold, C.A., Tiongo, M.M. (2003). Policy, Technical, and Environmental determinants and determination of the scaling up of livestock production in four fast growing developing countries; A synthesis. http://www.fao.org/ag/againfo/en/pplpi/docarc/LL12.pdf (accessed July, 2009).

Kataktalware, M.A., Nazar, S., Letha Devi G. and Ramesha, K.P. (2017). Commercial dairy production in India- present status, opportunities and challenges. In: Training Manual of National Training Programme on Commercial Dairy Production for Technical Staff of the ICAR Institutes organized at ICAR-NDRI, SRS, Bengaluru during February 27 to March 04, 2017. Pp: 1-7.

❑❑❑

Chapter-5

Ways to improve small farmer production system in India : A few thoughts

N.K.S. GOWDA[1] AND K. GIRIDHAR[2]

[1,2] Principal Scientist, ICAR- NIANP, Bengaluru

Small farmer production systems is a low input, resource limited system providing livelihood and ensuring sustenance of rural life. In view of its centrality in ameliorating poverty and providing food security, efforts have been made to improve this system to make it profitable. With the changing market trends and needs of consumers, farmers have to diversify their agriculture by integrating with livestock, poultry, horticulture and floriculture. Governmental policy support must include hassle free soft credit options, need based and time framed subsidies, reformed revenue administration, remunerative price for the produce without middlemen, crop insurance, long term measures against drought/floods and region specific agriculture/ livestock policies. Infrastructure development by way of road, electricity, communication, education, storage and marketing networks will have both direct and indirect effects in making the system more efficient. Institutional support should include skill and capacity development, timely advisories, precise forecasting and quality services. Resource sharing mechanism is vital for better small farming that includes cooperative societies, farmers clubs, self help groups, watershed committees and custom hire services of farm machinery. Research and extension agencies must focus on solving the existing problems and evolve novel tools to translate discovery and innovation to adoptable technologies. Productivity improvement under small farmers system needs integration of stakeholders, Government policies, Organizational support, Community participation and functional linkage between research and extension agencies.

Small farmer production systems (SFPS) occupying 45% of cultivable land, contribute to over 50% of agriculture production in India. This is a low input, resource limited system providing livelihood and ensuring sustenance of rural life. In view of its centrality in ameliorating poverty and providing food security, efforts have been made to improve this system to make it profitable. To achieve this, it is imperative to overcome the existing limitations. Over the years the marginal farmers have increased with 2.4 per cent compound annual growth rate, while number of large farmers decreased 2.5 per cent compound annual growth rate. For ensuring sustainable viability of marginal and small farmers, the creation of job opportunities in rural areas along with suitable policy support for development of livestock sector and other allied activities especially dairy, goat, sheep and backyard poultry farming would be the necessity for resource-poor farming community. In SFPS, the major limitations are access to fertile soil, water, marketing network, quality seeds, fertilizer and userfriendly credit facility. Measures to overcome these by empowerment in terms of land ownership, easy access to inputs, capacity building, social security, crop/livestock insurance and profitable rates for the produce will make the SFPS more sustainable and productive. Decline in the number of youths employed in agriculture due to migration to cities is a concern and hence need to be suitably addressed and make agriculture more remunerative. Certain strategies and policies that are required to improve the SFPS are briefed in this paper.

Awareness at Farmer Level

Small farmer production cannot be a standalone system. It must be integrated with other related systems like crop, livestock, poultry, horticulture, floriculture, sericulture etc. This interlinkage provides complementary inputs byway of fodder, manure and supplementary incomes like sale of milk, meat, fruits, flowers etc. This is an effective means of social security for small and marginal farmers. Judicious use of resources like water by way of conservation methods is very important. Less input oriented crops like millets, livestock like indigenous cattle, small ruminants, backyard poultry and piggery are the most suited option in SFPS. Farmer need to be aware of the Government schemes available and he should consult the field officers to get the inputs. Consulting the Department officials, fellow farm men or women and established model farmers in the region will create better awareness about farming, soil health, recent developments in the market and inturn the confidence level of individual farmers.

Role of Government and Policies

Government as a policy making body has a pivotal role in improving the SFPS. Adequate credit facility in the form of soft loans is required. Many farmers want long term loan period of 3 or five years instead of 1 year. This will enable them for easy repayment of loans. Quite often the subsidies and social welfare schemes are not very relevant to the farmers need. Many farmers demand quality services instead of subsidies. Moreover, subsidies should be need based and timely available. The procedure of subsidy and credit scheme implementation should be simplified. The NABARD must enhance the support share for small and marginal farmers. The revenue administration needs revamping to become more farmer friendly. The policy of remunerative price for the farmers produce is a long waited issue. Like price of milk at Cooperative milk union level, price of other agricultural produce also need to be regulated. If profitable price is given for their produce, no farmers will leave agriculture. Government need to invest heavily on water harvesting, rejuvenation of lakes, rivers, interlinking of rivers, and also measures for overcoming floods and better conservation of flood water. Drip irrigation, water conservation are to be made mandatory. Regions prone for either drought or flood is quite known in the country and well thought preventive management measures need to be strictly implemented. Small farmers and livestock are the major sufferers during natural calamities. Each state in the country should have high powered technical committee for both agriculture, livestock and fisheries to make focused implementable policies. Region and location specific policies for crop, livestock and fishery need to be formulated. Depending on the resources like soil nature, water availability, rain fall pattern and weather conditions, type of agriculture and livestock for that region need to be delineated. Transportation network and road connectivity is yet another sector that need attention of Government. Roads need to be laid connecting village to agriculture land for movement of bullock carts or vehicles and machinery. Availability of electricity and communication in rural areas need to be further improved.

Institutional Support

The Government Institutions, Research organizations, Educational Institutions and Krishi Vigyana Kendras have a greater responsibility to address the problem of farmers. Promoting capacity building and skill development of farmers in the specific sector of crop, livestock or fishery should be one of the Institutional activities. Timely advisories on type of seed, soil amendments, type of livestock breed, feeding etc. will help the farmer to make his occupation more sustainable and profitable. Farmers under small production system dependent on services like fertilizer availability,insecticides,pesticides,artificial insemination, deworming, vaccination etc. The Departmental mechanism to deliver these services needs to be streamlined and effectively accounted and monitored. The Research Institutions should focus on existing problems of farmers and ways to solve them. Scientists should evolve novel technologies which are innovative and farmer friendly. Many a times, farmers will be innovative and give better ideas for research. Such farmers should be identified and involve in research advisory committees of Institutions. Innovative farmers day and honoring the innovative farmers for their accomplishments should be a regular activity of Institutions. Technology validation and testing must involve the end users like individual farmers, milk societies to get the perception about the technology. There are successful models of technology validation followed by immediate adoption.

For example - Area Specific Mineral Mixture (ASMM) technology of ICAR-NIANP, Bengaluru was field tested in selected milk societies of Karnataka Milk Federation (KMF). This technology was perceived very well by the farmers, and was adopted on a large scale by the KMF. Now the ASMM packets are available for dairy farmers in each society throughout Karnataka and this helped to overcome the mineral deficiency in dairy animals. Similarly, the areca sheath as an alternate dry fodder source was evaluated and field tested by establishing a model unit at Panaje milk society in Dakshina Kannada milk union. The perception and results of this technology was highly encouraging and now this model unit has become a working cum demonstration unit. With the support of NABARD, the Dakshina Kannada milk union has established 75 more such areca sheath processing units to solve the problem of dry fodder shortage in this region. Similar approaches have been adopted by scientists of NIANP to popularize fodder trees, azolla cultivation, green fodder cultivation. Now the trained farmers are the resource persons for azolla culture and fodder root slips to other farmers. In this approach horizontal extension is possible and is more practical under SFPS. The scientists and extension officials should contribute to radio and television programs to reach the farmers and disseminate technical informations. Information centers with dedicated phone and net connectivity are to be established in all Institutes. Successful interventions should be documented and published in news papers, magazines and broadcasted in electronic media.

Community Level Participation

Community level resource sharing mechanism is vital for small farming that includes cooperative societies, farmers clubs, self help groups, watershed committees and custom hire services of farm machinery. Farmers under small holding system need to adopt the concept of collective hiring, transportation and marketing. This will avoid the role of middlemen. There are very good models under this concept like community milking machines, community fodder plots and chaff cutters and village level water harvesting tanks.

Role of KVK and NGOs

Krishi Vigyana Kendras are the face of ICAR and links research, education and farmer. They need to be strengthened in terms of manpower, specialization and facilities to broaden the network of activity. There are reputed Non Governmental Organizations (NGO) who are doing excellent farmer oriented activities. They need to be supported for joint working with ICAR and State Universities/ Departments.

Summary

Productivity improvement under small farmers production system needs integration of small farmers, government policies, organizational support, community participation and functional linkage between research and extension agencies. Concept of horizontal extension by way of creating model farmers, working cum demonstration units at milk society and panchayat levels should be promoted. Mass media and communication has a great role to play in creating knowledge awareness. Recognizing and honoring a farmer's effort and achievement will help to create many more such successful farmers.

❑❑❑

Chapter-6

Small farmer production systems for southern Karnataka region

C. VASUDEVAPPA

Vice Chancellor, UAHS, Shivamogga, Karnataka

Global agricultural output must have to double by 2050 to meet expected demand of food, feed, fibre and fuel need of ever increasing population. Food production doubled during past 25 years, which has helped to reduce number of hungry people from 994 million in 1990-92 to 791 million in 2012-14 in developing countries.

Despite rapid economic growth trajectory of 8 percent, 1/4th of world's hungry and poor have their homes in India. Glacial progress in nutrition; over 40 percent of world's malnourished children are our own. Nearly 1/3rd of world's smallholder farmers are in India, comprising 50 percent of rural poor and hungry. India is a nation of youth, 65 percent of population being less than 35 years age despite formidable scientific manpower and knowledge pool, India has largest bulge of unemployed/poorly employed youth in world and is a demographic disaster.

India has over 17 percent of world population with only 2.4 percent of land area and 4 percentof water. Land, water, biodiversity will shrink by 30 to 50 percent; posing systems at risk. Distorted trade-offs amongfood security, ecological intensification and diversification, intensifying volatilities of climate change, market and food prices, entrenched high disparity breed inequality and blunt growth efforts.

Although Government of India is committed to sustain 4 percent growth in agricultural GDP focusing a technology driven second green revolution with emphasis on higher productivity, including a "Protein Revolution" and expanded "Blue Revolution" of inland fisheries.

Integrated Farming System – A Saviour for Small Holder Farmers

At present, farmers concentrate mainly on crop production which is subjected to a high degree of uncertainty in income and employment to farmers. In this context, it is imperative to evolve suitable strategy for augmenting income of a farm. Integration of various agricultural enterprises such as cropping, animal husbandry, fishery, forestry, bee keeping and mushroom production have great potentiality in agricultural economy. These enterprises not only supplement income of farmer but also help in increasing family labour employment.

In agriculture, apart from routine crop production, live stock production is another potential area and is a source of high quality protein, regular income (a sort of moving bank), gender empowerment and employment potential of almost 18 million directly employed in livestock sub-sector. Livestock must be driver of climate smart agriculture equipped with food safety and environmental regulations. This could become a driver of congruence of food, nutrition, health, additional income and livelihood security.

In this regard, interlinking of various adoptable enterprises definitely plays a key role, not only in meeting food demand of farm but also in discouraging dependence on external agri-inputs, providing year round employment by maintaining diversity in farm. Some of important IFS modules which can be practiced under different agro-ecosystems by small farm holders are presented below.

Dry-land

In areas of low rainfall, a farm unit of one hectare can profitably adopt cropping activity of climate resilient minor millets (finger millet/ foxtail millet/little millet) based oilseeds or pulse intercropping in 4:1 or 8:2 in 0.4 ha area, along with this, seed production activity in onion is found to be feasible to harvest maximum monetary returns in small area of 0.1 ha. In addition to create complementarities, dry land farmers can very well maintain 20 units of telicherry goat (19 female+1 male), 50 back yard poultry and two milching cows with compost production unit. To meet out input demand of said enterprises bund planting of high yielding fodder crops such as anjan/rhodes/ stylosanthus spp., perennial leguminous fodder trees viz., subabul, casurina and agri-horti system (high density sweet orange and groundnut) in 0.3 ha can be practiced to assure fodder and nutritional security. A farm pond of 250 m^3 for safe storage and recycling of harvested runoff is required. Remaining 0.2 ha of land can be used for kitchen gardening and to accommodate household activity, dairy, compost production and bio gas production unit with a capacity of 2 m^3/day. The above model would be economically feasible with net returns of ₹ 4,24,500/ha at BC ratio of 1:4.05 by employing 886 man days/year/ha with efficient recycling of all resources.

Irrigated Land

Under assured irrigation, rice-pulse or oil seed sequence cropping and areca nut + cocoa/banana+pepper multi-storied cropping activity, in an area of 0.3 ha each coupled with six cross bred milch animals, 50 back yard poultry with 2m^3/daybio gas production unit with appropriate size of compost production unit for effective recycling of recourses besides 2 kg/day mushroom production with 500 m^3 capacity farm pond to ensure one tonne of fish production in a season (0.10 ha) can be practiced. In addition to this, silvi-pastural system in an area of 0.10 ha is found most appropriate. The said model is very efficient in securing livelihood of small holder irrigated farmer with BC ratio of 4.90 by realising net return of ₹ 10,29,755/ha on a sustainable basis by providing year round employment to members of family.

Garden Land

In areas of garden land, perennial minor fruit crop based (sapota/ coorg mandarin/ram butan/wood apple/lichi etc) fodder production (horti-silvi pasture system) as cropping component in an area of 0.60 ha with three units of milching cows and cage rearing of 50 poultry birds with 20 units of sheep (19 female +1 male) by establishing appropriate sized compost production units and silage units with mushroom production unit @ 2 kg/day along with nursery activity on pepper seedling production and placing of 8 bee hives in garden, establishment bio-gas production unit with gas production capacity of 2m^3/day and 500 m^3 water storage capacity farm pond to ensure one tonne of fish production in a season (0.10 ha) are found best for garden land small holder farmer. The said model is economically viable and sustainable with BC ratio of 4.43 and net returns of ₹ 6,60,000/ ha by efficient recycling of resources with year round employment generation of 1002 man days/ha.

Coastal Land

In coastal region, cashew based pastural system and rice cum fish based pulse sequence cropping system in an area of 0.60 ha, fodder production with high yielding forage crops in 0.20 ha, 5 unit of milching cows, cage feeding of 50 poultry birds, stall fed goatery unit (19 female + 1 male), and nursery activity of cashew and pepper seedling production, bee keeping and 2 m^3/day bio gas production along with 500 m^3 capacity farm pond to ensure one tonne of fish production in a season (0.10 ha) was found best for costal land small holders. The proposed model for coastal land is economically viable with higher BC ratio of 4.01 by earning net returns of ₹ 6,34,500/ha with employment generation of 1390 man days/ha.

Case Study at AHRS, Kathalagere

- Integration of crop husbandry and allied enterprises such as dairying, horticulture, sheep rearing and fishery was practiced since three years. Adoption of IFS module resulted in a net profit of ₹ 1,70,580/- annually with a BC ratio of 3.76 for an average income of ₹ 14,170/month.
- The input generated and recycled was monetary worth of 52,889/- from various inputs such as crop residue, dung urine and other farm wastes. This can generate nutrients worth of 363 kg for an equivalent market value of 5110/- (including NPK and Micronutrient).
- A total of 1070 man days per year employment was generated.
- An efficient utilization of one component output as an input of other component of farming was realised.
- A family having an average of five members can effectively get engaged in various enterprises throughout year.
- Thus, adoption IFS module will enhance socio-economic status apart from providing food and nutritional security on a sustainable basis.

Summary

This paper examines role and challenges of small holding agriculture. India has done well in attainting food grains self sufficiency since days of green revolution. Difference between green revolution period and expected 'second green revolution' is that globalization challenges, volatility in prices, shrinking farm size, rain fed farming challenges and environmental stress and climate change. Small farmers are certainly going to remain in India for next decade or more.

Way Ahead

- Main goals are raising agricultural growth productivity, move towards high value agriculture and rural non-farm and maintain food security, shared growth (small farmers, lagging regions, women etc. and sustainable growth.
- There are six deficits in agriculture. These are: investment, credit, infrastructure, research, extension (technology), market, diversification, institutions and education/skill.
- Food security has been an overriding priority in India since 1960s. focus paid off with attaining food grain security and poverty reduction.
- The future growth will need to rely on a dual strategy of diversification into climate resilient food crops and rainfed areas with focus on dryland hortuclture, agro-forestry, sheep and goat rearing.
- Basically investment and institution strategies have to focus on small and marginal farmers, women farmers, rainfed areas, environmental stress, rural non-farm sector for sustainable agri-transformation.
- Small farmers have difficulties to access inputs, credit and extension or to market their output. Hence, all inputs are to be produced internally on farm.
- Investment in infrastructure like roads, irrigation, communications etc. can benefit small farmers.
- Many institutional innovations are coming up to link small farmers to high value agriculture and help increasing their productivity and marketing (contact farming, dairy sector, cooperative marketing).
- ICT revolution in India is also helping small farmers with information, input supplies and marketing.

- Risk mitigation strategies are needed for small farms.
- There is a recognition that small farmers need a special focus in order to have growth and equity in agriculture and food security.
- Small farmers need agricultural diversification and income from rural non-farm sector.
- Need for massive increase in investments for agriculture and rural infrastructure by simultaneously improving institutions for better delivery systems.
- Such a big push is needed for agriculture sector for food security, livelihoods, sustainability and transformation.

❑❑❑

Chapter-7

Small farmer production systems : KVK Kodagu experiences

SAJU GEORGE[1] AND B. PRABHAKAR[2]

Senior Scientist and Head[1] and SMS (Horticulture)[2]
ICAR- KVK (IIHR, Bengaluru), Gonikoppal, Kodagu-571231, Karnataka

Kodagu district in Karnataka is famous for its coffee based cropping system. Some of challenges which small farmer production systems may face in future in Kodagu district are loss of biodiversity due to tourism expansion, pollution, population and infrastructure pressure, rapid depletion of water tables etc. Small farmers need to use available land innovatively with a proper understanding of growing needs and demands of consumers. There are many issues facing small farmer production system in district like lack of bargaining power, lack of knowledge of new technologies, high cost of inputs used in agriculture etc. This paper highlights some of issues which emerged during focussed group discussions conducted by Krishi Vigyan Kendra with farmers in different talukas of district. KVK Kodagu has taken up many on campus and off campus training programme, farmers meet, technology weeks, field days, advisory services through print and electronic media to strengthen small farmers production capacity through technological interventions looking into these issues.

Horticulture plays an important role in rural economy of Kodagu district. The district is famous for its coffee-pepper based cropping system. Coffee is major crop in district, which account for around 75% of gross cropped area. Paddy is major cereal crop of district. Black pepper, banana, arecanut, ginger and coconut are other important horticultural crops of area.Since two major commodities of district-coffee and pepper are exported, apart from meeting domestic demand, open market operations influenced by international demand, has thrown up several challenges for small farmer production systems in district. This paper focuses on bringing some of issues faced by small farmers of this hilly district to fore, along with measures to overcome these challenges.

Methodology

Six Focussed Group Discussions (FGD) were conducted in all 3 talukas to identify small farmer's problems and participatory solutions. Based on focussed group discussions carried out, some of major issues identified have been broadly categorized into major issues along with way forward in consultation with farmers. Farmers were asked to rank these issues in order of their priority.

Results

Some of challenges which small farmer production systems can face in future in Kodagu district are loss of Biodiversity due to tourism expansion, pollution, population and infrastructure pressure, rapid depletion of water tables etc. Kodagu district has 68454 total land holdings operating 157510 ha of area. Marginal and small farms account for 41 and 25 percent of holdings and operate 9 % and 15 % of total area in district (Table 1).

Table 1. Number of Land holdings and Area operated by Class Size in Kodagu district

S.No.	Farm Size	No.	% Share	Area	% Share
1.	Marginal	28101	41.05	13991	8.88
2.	Small	16945	24.75	23777	15.10
3.	Semi- Medium	13455	19.66	36771	23.35
4.	Medium	7787	11.38	45116	28.64
5.	Large	2166	3.16	37853	24.03
	Total	68454	100.00	157510	100

(Marginal- < 1 ha, Small: 1-2 ha, Semi Medium: 2-4 ha, Medium: 4-10 ha, Large: > 10 ha) (*Source:* C-DAP, Kodagu, 2016)

Issues and problems faced by small farmers in hilly ecosystems

1. **Lack of bargaining power in market:** Small farmers always lack bargaining power and often they have to sell their price produce, at throwaway prices dictated by middleman. The same is case when buying inputs for their agricultural operations.

2. **Lack of technological information with small farmers:** Small farmers need more technological information. It is often seen that small farmers get agricultural technological information either late or even when they hear about new technology they may not be convinced enough to adopt new technologies, as that may entail risks. So, mechanisms need to be worked out so that small farmers are involved in technology testing, evaluation, at demonstration farms of KVKs or other Agricultural Institutes. This will build confidence of small farmers in power of technologies to transform their lives. This requires concerted efforts from

Government to strengthen existing KVKs, so that their capacity to reach more no. of small farmers is enhanced and thus, small farmers can be empowered technologically and thereby economically. This will go a long way in meeting stated objectives of Government to double farmers income by 2022.

Way Forward

1. **Formation of farmers collectives:** To mitigate such problems way forward is to form farmer's collectives. These collectives often have volume of produce to attract genuine buyers, who can offer them maximum price for their produce. The same is case when going for purchase of inputs, farmer group can always get inputs at most competitive prices if they can pool their requirements and place order directly to manufacturer or main dealer of inputs like fertilisers, pesticides, bio-inputs etc. This has been amply amplified by farmer produces organisations created or in process of creation in Kodagu district. One of FPO is called Bhragandeshwara farmers producers company which is promoted by Department of Horticulture in Madikeri Taluk and has more than 1450 farmer members. They have started pooling their produce for collective selling. They are now able to fetch

higher marketing price for their produce like coffee, ginger, honey etc. To solve problem of low prices of coconut, they have decided to go in for value addition like coconut oil instead of selling raw coconuts. This has helped farmers in getting a better price and profit for farmers.

Similar is experience with another FPO called Puthari Farmers Producers company being promoted by KVK, Kodagu, which is under process of registration and is expected to get financial assistance from NABARD. When they started pooling requirements of agricultural inputs like agricultural lime, dolomite, shadenets, agricultural irrigation pipes, pesticides, they were able to get them directly from producers at far cheaper rates, than retail prices of produce in market. Shortly they have plans to go for selling their main produce like coffee and pepper, through branding and selling directly to exporters. Since Kodagu coffee and pepper are grown in shade without much damage to environment and maintaining diversity, they are confidant of fetching a higher price in market. The importance of formation of farmer collectives in increasing farmers income has been amply proved in case of Mahagrapes (Roy and Thorat, 2008).

2. **Role of new age ICT technologies like mobiles:** Mobiles are now a days playing an important role in transforming India and they may have a greater role in transforming rural India.. Mobiles are now ubiquitous. This can be used as on effective tool in dissemination of new technologies, information, market prices etc. KVK Kodagu is sending messages regularly to more than 4000 farmers through Kisan Mobile advisory on subjects like fertilizer application, pest management, events of agricultural importance etc. Messages are in form of SMS. After reading messages, farmers can always revert back to KVK either through telephone or in person for more information. This process keep farmers interest in agricultural alive, as they feel important and empowered as public extension agencies are contacting them regularly. Since most of these farmers are key farmers in their areas of operation, information further spreads through lateral diffusion to other farmers in district.Diffusion strategies, as currently practiced by most change agencies, often lead to increased inequity and therefore might be used as guides for devising revised strategies which avoid what currently practiced strategies predict. This view implies that those who seek more

equitable development programs should not carry out further surveys of current diffusion practice, but rather experiment to develop and test new alternatives to such usual diffusion programs (Roling and Chege, 1976).

3. **Integrated Farming System (IFS):** It is often seen that farmers practicing IFS are most insulated from distress situation that has become more common in recent past. IFS through its integrated approach, provides means for farmers to get inputs like Farm Yard Manure at lower cost, a regular source of income through animal products like meat eggs, etc., round year. Farmers are assured of income from animal components, as agricultural produce is often seasonal and there is wide variations in their prices also. So, developmental agencies working in rural areas have to promote IFS to strengthen small farmers in district.

4. **Pradhan Mantrhi Phasal Bheema Yojana** (PMFBY): Insurance Schemes for farmers are important, as farmers regularly face climate vagaries like monsoons, draught situations etc. The Government has come up with a scheme called PMFBY, where farmers contribution is very low. Therefore more farmers need to be made aware of scheme, so that their enrolment can go up. All these measure can go a long way in protecting farmers interests and strengthening them economically and technologically.

Summary

The interaction with small farmers have thrown out many challenges being faced by them. But there are appropriate tools and mechanisms and technologies to overcome those issues. These are formation farmers collectives, providing timely technical knowledge to farmers, use of new age technologies like mobiles, integrated farming system and new government programmes like PMFBY etc. The way forward for strengthening small farm holders in hilly ecosystem is to encourage these small famers to go for making farmers collectives, which can bring convergence of many Government programmes at a single point of delivery, thus benefitting maximum farmers.

Reference

Anonymous, (2016). *Comprehensive District Agricultural Plan, Kodagu.* Published by Kodagu District administration.

Röling, N.G., Ascroft, J., & Chege, F. W. (1976). The diffusion of innovations and issue of equity in rural development. *Communication Research*. 3(2): 155-170.

Roy, D., & Thorat, A. (2008). Success in high value horticultural export markets for small farmers: The case of Mahagrapes in India. *World Development*. 36(10): 1874-1890.

□□□

Chapter-8

Efficiency enhancement of small farmer production systems: Experiences of KVKs in southern states of India

CHANDRE GOWDA M.J[1]., AND SREENATH DIXIT[2]

[1]Principal Scientist and [2]Director, ICAR-ATARI, Bengaluru

---

A study of successful smallholder farmers in Karnataka, Kerala and Tamil Nadu was carried out during 2016. The data indicated that 50.8 percent of farms adopted integrated farming system and equal number of households (24.6% each) were crop-based and enterprise oriented. Analysis of components of households revealed that 80.7 percent of all households cultivated horticulture crops whereas 63.2 percent households reared a livestock component. Annual income of crop-based farms was ₹ 1.72 lakh compared to ₹ 2.74 lakh in integrated farms and ₹ 3.52 lakh in enterprise-oriented farms, although difference was not significant. About 20% of households were headed by women, out of which 27% practiced enterprise-oriented farming system. Income per unit of land was significantly higher among women headed households. Income per acre was also significantly higher in enterprise oriented farms over other two categories. Annual income was higher by ₹ 14000 in crop-based farms belonging to women-headed households. It was higher by ₹ 21000 in integrated farms and by ₹ 78000 in case of enterprise oriented farms of women as compared to those headed by men.

Smallholder agriculture is practised by families (including one or more households) using only or mostly family labour and deriving from that work a large but variable share of their income, in kind or in cash. The definition of "smallholder agriculture" varies with specific context at regional, national and local levels, and also over time as economies transform. A smallholding is "small" not necessarily due to size of holding but because resources are scarce, especially land, and using it to generate a level of income that helps fulfil basic needs and achieve a sustainable livelihood consequently require a high level of total factor productivity, requiring in turn a significant level of investment (HLPE, 2013). Smallholder agriculture plays a major role in national economy of many countries, particularly in India as 67.1% of holdings are less than 1 ha and 17.9% of holdings are 1-2 ha, thus constituting 85% of total holdings (GoI,2015). As agricultural growth and overall development are strongly inter-linked in countries where a smallholder agriculture dominates, it is important to give adequate emphasis for development of smallholders in India. Two important dimensions related to smallholders demand maximum attention in present scenario, which is dominated by declining profits and increasing climate variabilities. First, small farmers play an essential role in in-situ preservation of biodiversity, and secondly, women small farmers have developed a system of community conservation of biodiversity through in situ on-farm conservation (Swaminathan, 2010). Development of smallholder agriculture resulting in successful smallholders agri-business will serve to motivate other farmers through farmer-to-farmer multiplier effect. The factors that contribute to their success have to be understood and replicated with necessary policy and governance support. It has been realized that agricultural technologies have played crucial role in enabling farmers to achieve sustained development. The Indian Council of Agricultural Research has a network of research institutes and agricultural universities to generate technologies suitable to diverse agro-climatic regions. In order to facilitate technology delivery to potential end users in all agro-climatic situations, ICAR has also established Krishi Vigyan Kendras (Agricultural Science Centres) at district level to serve as frontline extension mechanism, which link research institutes to development departments and farmers on location-specific basis. Smallholders supported by Krishi Vigyan Kendras (KVKs) provide constant feedback on technological, social, economic and policy support required for development in a sustained manner. This study is an attempt to understand efficiency enhancement dimensions of successful smallholders.

Methodology

Three South Indian States namely Karnataka, Kerala and Tamil Nadu have at present 31, 14 and 30 KVKs respectively, covering almost every district in three states. Eleven KVKs were randomly identified from three states and were asked to provide list of smallholders who frequented KVKs, participated in at least one programme of KVK in 2016 and who considered themselves successful in their farming. Fifty-seven successful smallholders identified by KVKs were contacted with a semi-structured interview schedule to generate data of their farm activities, variable costs, gross returns and factors that contributed to success in farming. Smallholders were categorised, based on their chief income generating activity, into crop-based, integrated farming-based and enterprise-oriented families. Enterprises oriented households practiced income generating activities like apiculture, mushroom cultivation, value addition, seed production etc., in addition to crop and livestock activities.

Results

Half of smallholders (50.8%) practiced integrated farming, a combination of crops and livestock in their farms (Fig.1). About a quarter of respondents each practiced crop-based farms, mostly horticulture crops like vegetables and fruits. The other quarter of respondents practiced enterprise-oriented farming, which is more of specialised within an integrated farming system.

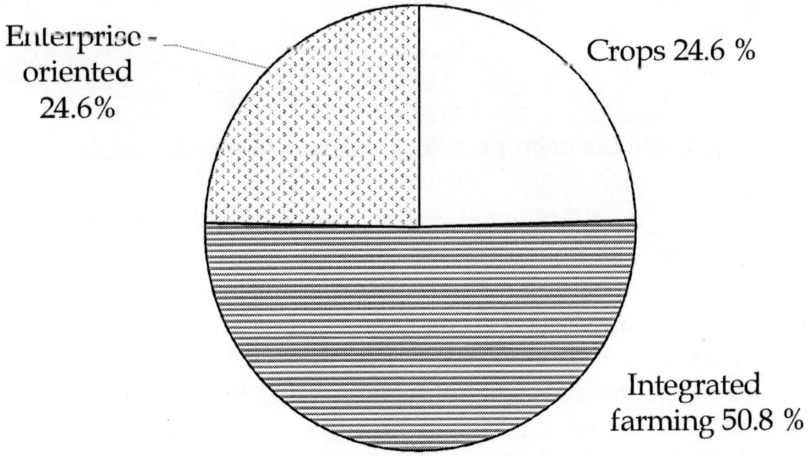

Fig. 1. Choice of Crop and Enterprises of Successful smallholders

There was no difference in average landholding of three categories of successful smallholders. The enterprise oriented farms were slightly larger than smallholders who practiced integrated and crops-based farming. However, difference was statistically not significant. The limited land available with smallholders, makes them to adopt wide-variety of crops and activities to sustain family livelihood. As evident from details in Fig. 2, horticulture crops were adopted by 80.7% of farmers. Dairy, sheep and goat, poultry and fisheries constituted livestock activities and were part of majority of households. Horticulture and livestock are two important sectors that are maintaining higher growth rate and also larger contribution to agricultural GDP of country. Similar trend is also seen among smallholders in present study.

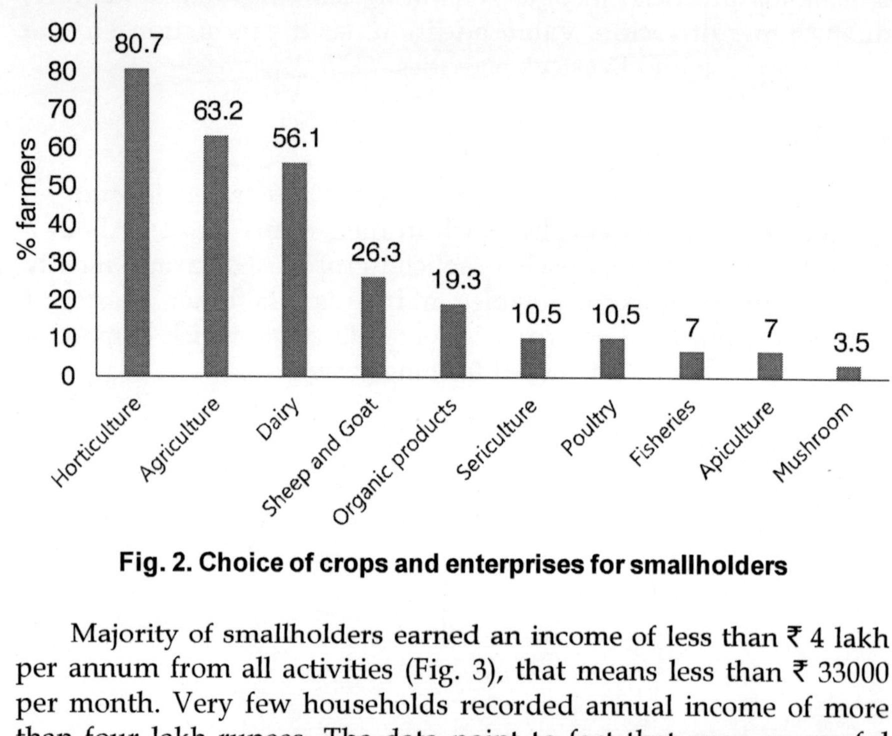

Fig. 2. Choice of crops and enterprises for smallholders

Majority of smallholders earned an income of less than ₹ 4 lakh per annum from all activities (Fig. 3), that means less than ₹ 33000 per month. Very few households recorded annual income of more than four lakh rupees. The data point to fact that even successful smallholders are struggling to generate high income from their farm-based activities. One of households, which had limited land (0.1 acre) earned more than four lakh per annum as it depended more on livestock activities and also ventured into processing and value addition. However, two households that generated more than Rs.10 lakh per annum had a farm size of about 5 acres, upper limit of smallholder land category.

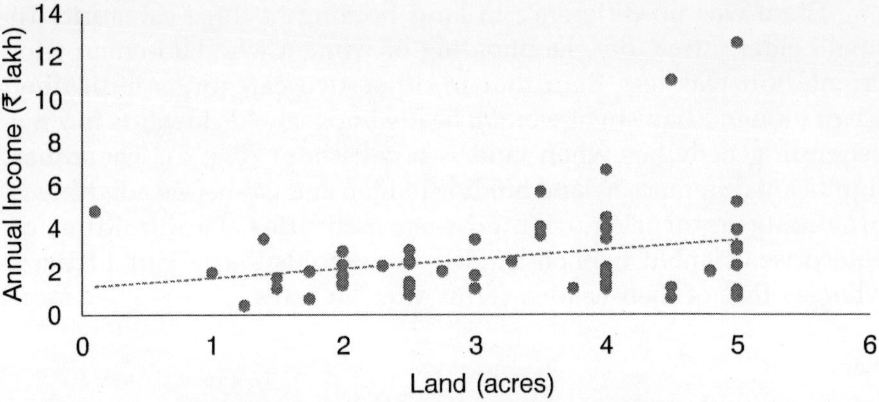

Fig. 3. Land and Income scatter plotting

Although there was no direct relationship between land holding and income among smallholders, there was significant difference in income of three-category of farms/households. Households that practiced only crop-husbandry earned just half income earned by enterprise-oriented farms (Fig. 4). Integrated farms stood in between crop-based and etnerprise-oriented farms with an average income of ₹ 2.7 lakh per annum. The data signifies imprtance of diversification (integrated farms) and specialization within diversification (enterprise-oriented farms) to achieve higher income compared to only crop-husbandry.

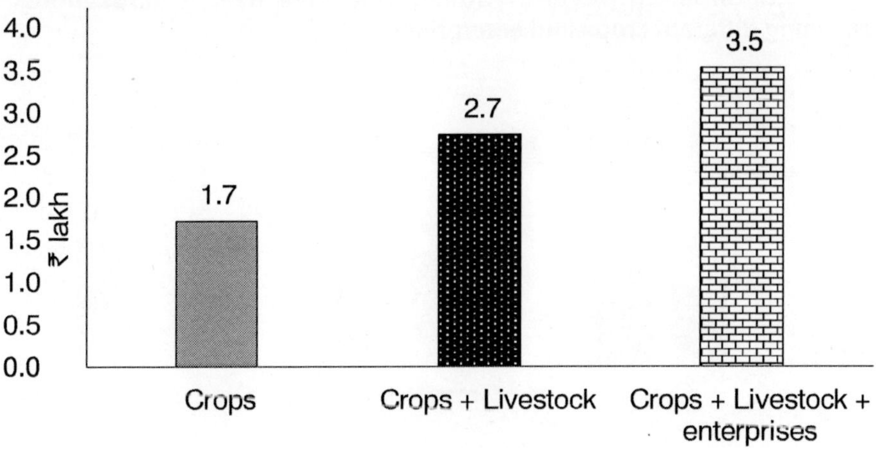

Fig. 4. Annual net income of households

There was no difference in land holding of three categories of smallholder farms. But, landholding of women who had enterprise-orientation was less than that in other two categories indicating adaptation mechanism of women headed households towards income generating activities when land is a constraint (Fig. 5). There was significant difference in land holding of men and women-headed farms practicing enterprise-oriented farm activities. Landholding of enterprise-oriented women-headed households was about 1.8 acre, whereas that of men-headed farms was 3.6 acres.

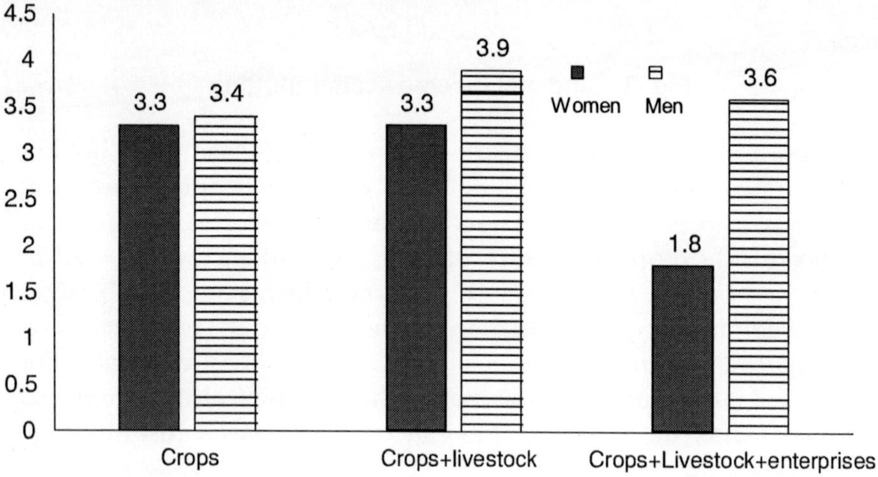

Fig. 5. Landholding (acre) of women and men headed households practicing different crops and enterprises

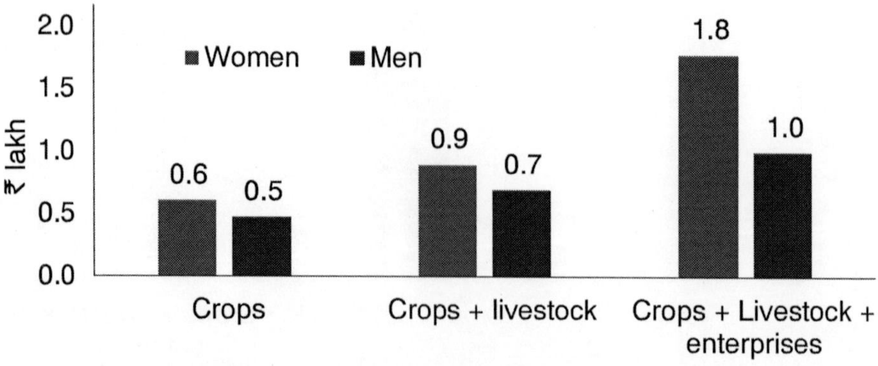

Fig. 6. Annual net income/acre for women and men headed households

When land was a constraint, women had specialized activities like processing and value addition, seed production, mushroom cultivation as a measure of income augmentation. Women-headed households generated ₹ 1.8 lakh per acre compared to ₹ 1 lakh per acre generated by men-headed farms (Fig. 6).

Factors Contributing to Success

Successful smallholders cited five important factors that contributed to successful farming (Table 1). The foremost was adoption of location-specific technology, by about 93% of households. This is exactly role played by Krishi Vigyan Kendras helping farmers to choose and adopt technologies that suit their farms, family resources and markets. Training and exposure visits to KVK or organized by KVK for other places, including successful farmers, was second most important factor for about 60% households. KVKs are knowledge and resource centres wherein entire farm is developed as instructional arena with demonstration units and learning centres. Farmers visiting KVKs not only learn from KVK but also get a mindset to experiment their own ideas in their farms. Sustaining income of smallholders demands not only integrated farming, but also some kind of specialization, which provide sizeable income. This is possible with activities like seed production, value addition etc. Marketing strategies are vital to ensure that their produce receives right price. About one-third of households attributed their success to innovative marketing, including direct marketing to consumers as one of reasons for success. Being part of farmers organizations was equally important. This process of collectivization is gaining momentum and will surely drive more success among smallholders.

Table 1. Factors contributing to successful smallholder farming

Rank	Factors	Farmers (%)
I	Adoption of location-specific Technology	92.9
II	Participate in Training and exposure visits organised by KVK	59.7
III	Specialization to income generating activity	35.1
IV	Innovative marketing strategies including direct marketing	31.6
IV	Group affiliation and being part of SHGs and FPOs	31.6

Summary

Integrated farming, diversification towards income generating activities and women-leadership are crucial efficiency enhancing mechanisms revealed in study. Smallholders must be supported and encouraged to identify suitable farm-specific enterprises to sustain family income. Smallholders must be educated, empowered and enabled with policy and governance support to think big in agri-business lines.

Way Forward

The present success achieved by smallholders is mostly restricted to production related outputs. Very few smallholders are practicing value addition and processing as an avenue to increase income. Scattered location of successful smallholders is coming in way of attaining critical mass required for transforming agriculture into agribusiness. Technology uptake and resource-use efficiency are directly related but not fully harnessed among smallholders due to their unorganized activities at all stages–pre-production, production and post-production stages. Collectivization, aggregation and integration in form of Farmer Producer Organization is a potential approach to provide much needed value-chain expansion.

Reference

HLPE. (2013). *Investing in smallholder agriculture for food security*. A report by High Level Panel of Experts on Food Security and Nutrition of Committee on World Food Security, Rome.

http://www.fao.org/fileadmin/user_upload/hlpe/hlpe_documents/HLPE_Reports/HLPE-Report-6_Investing_in_smallholder_agriculture.pdf

GOI. (2015). *Agricultural Statistics at a Glance 2014*. Government of India, Ministry of Agriculture Department of Agriculture & Cooperation, Directorate of Economics & Statistics, New Delhi.

□□□

Chapter-9

Group approaches for promoting sustainable smallholder farming

SREENATH DIXIT[1] AND M.J. CHANDRE GOWDA[2]
[1]Director and[2]Principal Scientist, ICAR-ATARI, Bengaluru

India lives in villages and rural livelihoods are largely dependent on agrarian activities. As per Agriculture Census 2010-11, total number of operational holdings in India is 138.35 million. Average size of operational holdings has further declined to 1.15 ha from 1.23 ha in 2005-06. Indian agrarian population consists of a large number of famers owning a small share of cultivable lands and a small number of farmers with a large share of total cultivable lands. About 85% of holdings belong to small and marginal farmers (below 2 ha) representing 45% of operated area. Whereas, medium sized operational holdings (2-10 ha) represent 14% farmers who own another 45% of land holdings. Large holder farmers who constitute just 1% of farmers own remaining 10% of holdings. Diverse nature of agriculture in India, with its wide variety of agro-climatic regions and broad range of socioeconomic conditions in rural population, calls for agricultural extension approaches that are context-and situation-specific. With about 85% of Indian farmers cultivating an area of 2 ha or less there is an increasing need for stronger intermediaries that can facilitate information access for diverse smallholder farmers.

Capacity of smallholder farmers to articulate their needs will influence their ability to obtain information they need. Considering large number of marginal and small land holdings in India, farmers' organizations (FOs) and self help groups (SHGs) play an important role in articulating needs of farmers to knowledge intermediaries. FOs and SHGs can be strengthened to operate side by side with either NGOs or public sector, but challenges exist in both sectors. Public capacity to build FOs and SHGs is limited, while NGOs with an understanding of these nuances are not numerous, rely on donor funds and would need public support to develop technical skills to facilitate groups (Swanson 2008; Sulaiman and Holt 2002). Within FOs or SHGs, problems related to social identity, including gender and caste, mean that these organizations may not be completely inclusive and are subject to elite capture problems. When farmer groups interact with others institutions, social identities and other social status perceptions mean that they may be too weak to articulate their concerns (Sulaiman and Holt 2002). Building capacity of such groups, and promoting development of leadership and management skills so that smallholder farmers can demand information they need, is therefore an important component of agricultural extension approaches. In this context, group approaches to smallholder farmers assume greater importance. Scaling-up of group approaches requires a professional understanding of groups in terms of their characteristics such as interaction, interdependence, structure, size, norms, roles, goals and cohesion are core competencies required by extension professionals (Suvedi and Kaplowitz, 2016). This paper, therefore, attempts to bring out and share certain examples from field where group approaches were successfully tried to enable smallholder farmers in achieving sustainable production systems.

1. Networking of bore well for Operationalizing Protective Irrigation

There is ample research evidence to show how rain-fed crops will benefit greatly if they are supplied with life-saving irrigation at times of mid-season droughts. However, there are not many instances where it was put into practice in real time under farmer's field conditions owing to many operational constraints. Protective irrigation is often a concept that was limited to research farms or seed farms, where irrigation was provided when moisture stress was encountered during cropping cycle. Under National Agricultural Innovation Project (NAIP) of Indian Council of Agricultural Research, providing protective irrigation was tried on a large scale by adopting social cum irrigation engineering techniques.

Malkaipet Tanda, a tribal hamlet located in Parigi Mandal of Rangareddy district, (now in Telangana State) was chosen for implementation of concept of water demand management through sharing of groundwater resources. During initial discussion with farming community it was found that 18 farmers shared their *pattas* (land records)on a 45-acre stretch, which had three functioning bore wells and one that was defunct. Of these 45 acres, only seven acres of paddy was being cultivated, while remaining land was kept fallow during *rabi* season owing to lack of irrigation source. A series of consultations were held with bore well owning farmers and neighbouring ones who had no water source to irrigate their fields. Year long negotiations with community to implement social regulations for groundwater usage finally yielded results. Over 45 acres of land belonging to 18 households was brought under a protective irrigation regimen by laying out a network of pipelines connecting three bore wells. Entire group of farmers agreed not to cultivatepaddy during *rabi* season but to share bore well water among themselves for growing irrigated dry crops.

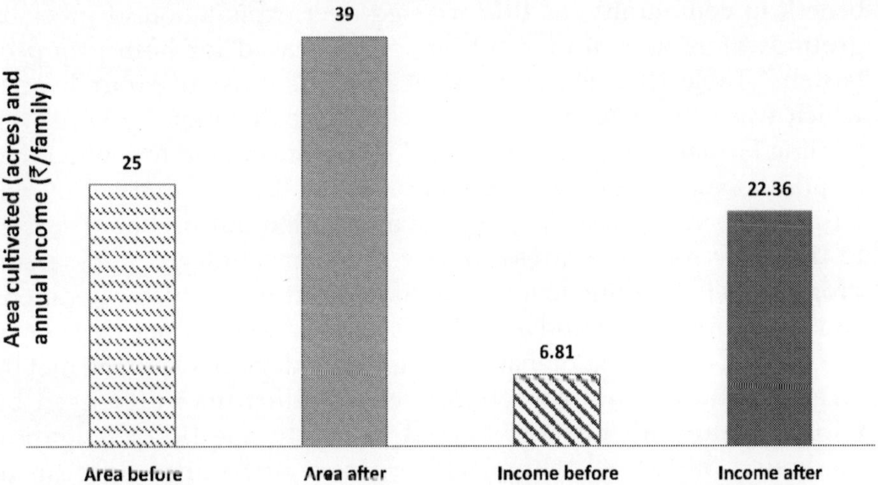

Fig. 1. Change in area cultivated and annual income of families without borewell due to network with borewell owners

Through this innovation, which involved irrigation and social engineering approaches, concept of protective irrigation was operationalized under field conditions. Comparing their cropping and income assessed impact of groundwater sharing on 'pipeline network user group' during year 2008-09 (when pipeline network was not operational) with those in 2009-10. Summary of impact is presented in Table 1. There was quantum jump in annual income of both groundwater donors and recipients offering a win-win situation for both. Once underground pipeline network was laid out to link bore wells, area under second crop increased by two folds. Eleven farmers who had never cultivated (crops during post-monsoon season (*rabi*) (could significantly increase their family (income thereby bridging gap between them and bore owning farmers (Fig. 1).

Further, they were able to provide (protective irrigation and save crop during (mid-season droughts besides diversifying (their cropping pattern during *kharif.* (Pipeline network encouraged them to stay (back and engage in cropping during post (monsoon season without migrating from their village, as this new system (generated an additional 90 to 100 days of (employment per family in post-rainy season. (Pipeline networking also prevented digging more bore (wells in village competing for scarce groundwater. Thus, pipeline network helped (to manage water demand within threshold of seasonal hydrological recharge potential. This is a very significant ecological benefit to community, as this arrested over exploitation of precious groundwater. Area under cultivation increased for both group of farmers (Table 1). This also resulted in efficient use of groundwater, which was otherwise used for only paddy cultivation by bore well owning farmers. Other farmers used water thus saved for cultivating an additional crop during post-rainy season. Besides, entire group of farmers was had access for protective irrigation during kharif season to tide over moisture stress in case crops encountered mid-season droughts. This arrangement was worked out over several iterations and by forming a groundwater users' association. As per by-laws of association, cost of maintenance of pumping devices would be met by farmers who did not own bore well in return to water used by them. Further, all farmers agreed to not to use flood irrigation methods. This also led then Government of Andhra Pradesh to adopting a policy of water sharing in their Indira Jala Prabha Yojana in which community bore wells were promoted instead of individual bore wells.

Table. Impact of groundwater sharing on cultivated area and income among user group

Before Groundwater Sharing Arrangement								Annual Income/family (₹)	After Groundwater Sharing Arrangement										Annual Income/family (₹)
	Area cultivated in Kharif (Acres)			Area cultivated in Rabi (Acres)			Total		Area cultivated in Kharif (Acres)						Area cultivated in Rabi (Acres)			Total	
	Paddy Red	gram + Maize	Total	Paddy	Ground nut	Total			Paddy	Redgram	Maize	Greengram	Cow pea		Paddy, ground	Ground nut	Total		
Bore well owning farmers (7 nos.)	7	0	7	2	10	12	19	-	5	2	0	7	2	16	5	7	18	27	-
Income of bore well owning farmers (₹'000)	49	0	49	14	60	74	123	17.6	32	6	0	25	6	69	70	42	112	181	26
Farmers without bore wells (11 nos.)	0	25	25	0	0	0	25	-	0	11	4	11	3	29	0	10	10	39	-
Income of farmers without bore wells (₹'000)	0	75	75	0	0	0	75	6.8	0	33	18	33	9	93	93	60	153	246	22

2. Promotion of on-farm generation of organic matter through vermicomposting

Maintaining or arresting decline in soil organic matter is most potent weapon in fighting against soil degradation and ensure sustainability of agriculture in dry-land areas. Consequences of depletion of organic matter are poor soil physical health, loss of favourable biology and occurrence of multiple nutrient deficiencies. It is stated that in rain-fed arid, semi-arid and sub-humid tracts, next to poor rainwater management; depletion of nutrients caused by organic matter deficiency is an important cause of soil degradation. Improving organic matter is, therefore, crucial in sustenance of soil quality and future agricultural productivity. A number of organic resources such as farmyard manure, crop residue, livestock dung and green leaf manures are available and these can be used to generate precious organic manure. Although vermicomposting is widely recommended as an effective method of converting degradable biomass, it is also fraught with several operational constraints at household level especially under rain-fed conditions. Keeping these in view, attempts were made to promote on-farm generation of organic matter through vermicomposting at community level under NAIP.

Backyard vermicomposting units promoted for individual households have been suffering with poor scalability due to its labour intensive nature. Therefore, there is a high rate of discontinuity of vermicomposting among smallholder households. In order to address scalability issues, project encouraged large-scale vermicomposting by farmer groups and women self help groups in a tribal hamlet Dupahad in Nalgonda district, Telangana and Pampanur village in Anantapur district, Andhra Pradesh. Training programs were conducted to train members of women self-help groups (SHGs) on technical aspects of breeding earthworms, management and collection of organic wastes, application of vermicomposting for various crops, packaging and sale of vermicomposting and bookkeeping.

a) Community vermicomposting cum biogas unit at Dupahad cluster, Nalgonda

Large sheds were erected for encouraging community vermicomposting. Trained women and youth were encouraged to prepare vermicomposting by using decomposable biomass and dung collected from village community. An arrangement was worked as per which smallholder farms were to cart semi/un-decomposed farm waste/biomass from their backyards to community vermicomposting units. These farmers would get fully decomposed vermicomposting in return to raw material supplied by them. Thus, vermicomposting was elevated to a specialized service providing enterprise from being a mere backyard activity. In addition to this, a community biogas unit was installed at community vermicomposting unit. Biogas unit used dung slurry for biogas production and passed on it for vermi-composting unit. Biogas unit was connected to a generator (15 KVA), which produced and supplied electricity to about 100 houses. These add-on features ensured viability of vermicomposting unit besides contributing to organic manure and energy production, besides clean environment. Additionally, 50 tribal households were provided with biogas connection for cooking purpose. This innovative arrangement ensured that all smallholder farms got benefit of turning their farm waste into valuable vermicomposting without having to deal with labor, space and water shortage, problems of storage and sale of vermicomposting and hassle of maintaining unit all by themselves. Besides, this arrangement could provide additional benefits to community such as electricity and biogas for cooking owing to scale of operation, saving on fuel and ensuring smoke-free rural kitchens.

b) Community vermicomposting cum calf rearing centre, Pampanur (Anantapur district)

A community vermicomposting initiative was taken up at Pampanur village, Anantapur district involving members of an SHG. A group of nine women were facilitated to initiate community vermicomposting. An old abandoned building belonging to a certain government department was identified and necessary permission was obtained from district authorities to use building to house vermicomposting and calf rearing units. With project support, building was repaired and nine vermicomposting units were set up under one roof.

Calf rearing units supported Vermicomposting unit and landless women of an SHG ran both units. A manger was constructed within building to stall feed calves. Initially nine crossbred calves were procured through revolving fund loan from project and later eight more calves were procured. dung from calves was used for vermicomposting. Azolla units were established and managed by women for fodder supplement. Project team negotiated with village temple authorities and got access for a piece of land in temple premises, which was used by women to cultivate fodder to support their calf rearing activity. This arrangement, while promoting complementing activities, also supported livelihood of women belonging to smallholder farms and landless families.

c) Farm pond

Farm ponds are dugouts that are excavated at vantage points on a farm to capture and store runoff so that rainwater thus collected can be used to provide life-saving moisture to crop during dry spells. Although farm ponds are being recommended as a means to offset mid-season droughts, smallholder farmers were particularly reluctant to adopt this technology until recently. There were many reasons for this. Some of prominent ones were apprehension of loosing precious land from their already smallholdings, lack of appropriate device to lift and use harvested rainwater and inadequate technical manpower to locate and recommend suitable site for and size of ponds. Hence, farm ponds to some extent remained confined to research farms and a few large holder farms and to a large extent, to research papers, articles and drought mitigation recommendations. Therefore, exclusive attention was paid under NAIP to bring this concept into practice by laying special emphasis on selection of site for digging farm ponds by considering principles of soil and water engineering. Further, care was taken to determine appropriate size of pond depending on crucial parameters such as rainfall and runoff pattern, topography and soil type.

Seethagondi village of Adilabad (Telangana) is endowed with fairly high rainfall (above 1000 mm) and deep black soils. Besides these, undulated topography in this area provides ideal opportunity for harvesting runoff, storing and reusing it for tiding over brief spells of drought during cropping season. Technical and economic feasibilities of runoff harvesting through farm ponds for profitable crop production and diversification were encouraging. Hence, to start with in 2008, an appropriate location was identified (after a detailed topo-survey of location) for a dugout pond (17m x 17 m x 4.5 m) involving a group of farmers as stakeholders. Initially there was very strong opposition from farmer Mr. Namdev and his brothers for allocating a part of their land for digging pond. By highlighting benefits of proposed intervention, farmers were persuaded to get farm pond dug in their land. Soon after farm pond was dug (July, 2008), there were good rains leading to complete filling of pond. They hired a diesel engine to irrigate 0.5 acre area where they grew tomatoes. By time tomatoes came to harvest, prices in market had touched close to 25/kg. Farmer earned a good profit of over 20,000/- through tomato cultivation. By mid November 2008, there was water still to a depth of two meters. This encouraged farmers to sow chickpea on one acre land. This raised enthusiasm (of farmers of this area and generally reluctant tribal farmers came forward in favour of digging farm ponds in their fields. Since demand for ponds was very high, project sought convergence with MNREGS for up scaling this intervention (See Box below).

Farm ponds: Up scaling with MNREGS

Success of farm pond in Adilabad generated a lot of interest among farmers as well as development departments in district. Mr. Namdev who was till then not known to many became a household name in surrounding villages. Several farmers who were reluctant initially, started approaching project staff with requests for farm ponds in their fields. This was due to a systematic awareness campaign undertaken by project involving district authorities and media. It included inviting key officials of development departments and encouraging them to arrange for farmer exposure visits to site of farm pond. This success story was widely shared with media and posted on project as well as ICAR websites and was shared during many discussions, meetings and seminars. This effort resulted in many more farmers showing willingness to adopt farm ponds on their fields. Taking advantage of changed attitude of farmers towards farm ponds, a detailed ground survey was carried out in surrounding villages of Seethagondi and a proposal was prepared identifying 30 suitable sites for farm ponds. Proposal was later submitted to nodal agency (District Water Management Agency; DWMA) that processed national rural employment guarantee scheme (MNREGS) works through Gram Panchayat. Proposal was closely monitored by project staff and community, which was favourably considered and 30 farm ponds worth 120 lakhs were approved by DWMA.

Summary

Examples cited above serve as very good evidence to show that it is possible to transform concepts of group approaches into actual practice under field conditions. All it needs is a multi-disciplinary and multi-institutional effort. Agricultural extension, or agricultural advisory services, comprises entire set of organizations that support people engaged in agricultural production and facilitate their efforts to solve problems; link to markets and other players in agricultural value chain; and obtain information, skills, and technologies to improve their livelihoods (Birner *et al.*, 2009; Davis 2009). This definition has evolved since T&V program, where focus of extension was transfer of technology to improve productivity, especially for staple food crops. While transfer of technology still has relevance, agricultural extension is now seen as playing a wider role by developing human and social capital, enhancing skills and knowledge for production and processing, facilitating access to markets and trade, organizing farmers and producer groups, and working with farmers toward sustainable natural resource management practices (Swanson, 2008). This is especially relevant while dealing with resource poor and marginalized smallholder farmers living in fragile ecosystems where group approach needs to be employed appropriately. Within this expanded role, breadth of information that agricultural extensioncan support-through provision and facilitating access and sharing-is much larger. In addition, as agriculture scenario has become more complex, farmers' access to sources of reliable and relevant information coupled with facilitation for application of such information in different farming situations has become increasingly important.

Reference

Birner, R., K.Davis, J.Pender, E.Nkonya, P.Anandayasekeram, J.Ekboir, A.Mbabu, D.Spielman, D.Horna, S.Benin, M.Cohn. (2009). From "best practice" to "best fit": A framework for analyzing pluralistic agricultural advisory services worldwide. *Journal of Agricultural Education and Extension.* 15(4): 341–355.

Davis, K. (2009). Agriculture and climate change: An agenda for negotiation in Copenhagen. important role of extension systems. *IFPRI Focus 16, Brief 11.* International Food Policy Research Institute.

Sulaiman, R. and G. Holt. (2002). Extension, poverty and vulnerability in India: Country study for Neuchatel Initiative. *Working Paper 154.* Overseas Development Institute, London.

Suvedi, M. and Kaplowitz, M. (2016). What every extension worker should know-Core competency handbook. *MEAS Handbook*, Michigan State University.

Swanson, B. (2008). *Global review of good agricultural extension and advisory service practices.* FAO.

□□□

Chapter-10

Group approaches for sustainable small farm production systems

A.K. SINGH[1] AND KESHAVA[2]

[1] DDG (Agricultural Extension), ICAR, New Delhi

[2] Principal Scientist (Agricultural Extension), ICAR, New Delhi

Indian agriculture is dominated by small and marginal farmers. As per Agriculture Census 2010-11[7], more than 85 percent of total operational holdings of 138.35 million are of less than 2 hectare area which account for 44.58 percent of total operated area. Thus, sustainability of agricultural production, productivity and profitability in country depend substantially on sustainable small farm production systems. Small and marginal farms can play an important role in India's food security and sustainable agricultural growth. However, farming on small farms, in general, is subsistence in nature with low value high volume crops and enterprises. The small holders are resource poor farmers and more vulnerable to natural calamities which are occurring frequently now a days. Increasing fragmentation of land holdings is a continuing cause for concern. Between 2000-01 and 2010-11, marginal holdings in country increased by 23 percent (from 75.41 million to 92.83 million) and small holdings increased by 9 percent (from 22.70 million to 24.78 million) (GoI, 2016).

The small holders in country are facing various socio-economic and structural challenges, some of them are unavailability of markets for inputs/products; absence of access to credit; poor human resource base; lack of access to suitable extension services; poor access to 'public goods' such as public irrigation, command area development, electricity grids; greater negative externalities from poor quality land and water management, etc. (Mahendra Dev, 2013).

Despite these challenges and poor conditions of small and marginal farmers, productivity and profitability of their farms can be improved by its diversification towards high value, low volume crops/ enterprises through technological advancements and innovative institutional arrangements. Several studies indicate that two components are very important for adoption of technologies: one is correct and timely information and knowledge to farmers; and second is availability of technological inputs such as seeds, fertilizers, etc. It needs change in focus of research and extension priorities and policy instruments.

Socio-economic Characteristics of Small Farm Production Systems

Small farms are characterized by subsistence farming with conventional production techniques and lack of mechanization. Generally, such farming families are engaged in cultivation of traditional crops and rearing of low productive animals for self-consumption. Since they are resource poor, young ones of families migrate to urban areas in search of livelihood and majority of elder ones are engaged in non-farm activities. The women along with some old family members are engaged in agriculture and allied activities.

Present Agricultural Scenario

Indian agriculture has travelled a long journey from an era of food shortage and import to that of self-sufficiency and export. With 252.23 million tons of food grain production during 2015-16 and all time high expected in 2017-18, India is World's largest producer of milk and among top 3 global producers of wheat, rice, cotton, fruits and vegetables. Citizens of country have witnessed green, white, yellow and blue revolutions. However, agriculture in India is facing issues like fragmentation of agricultural land holding, degradation and shrinkage of natural resources, pollution of underground water, deterioration of soil health, pesticide residue in agricultural produces, unavailability of quality inputs, increasing cost of inputs and decreasing profitability per unit of area. All these factors are causing concern for farmers, particularly small and marginal ones.

The field extension system in India has undergone many structural and functional changes since independence. However, certain limitations like ever-growing population, low manpower, inadequate budget for extension activities, administrative rigidity, old policies, narrow extension agent–farmer ratio of 1:5000 and involvement of extension personnel in non-extension jobs have made field agricultural extension system redundant (Davis *et al.*, 2010; and Mukherjee and Maity, 2015). Funding in extension is rich dividends and return to investments from extension is between 12 to 216 percent, average being 75 percent. Thus, investment in extension need to be further enhanced (Pal, 2017)

In changing agricultural scenario, farmers need different sets of information and services like credit availability, quality inputs, and post-harvest management like processing, value addition, storage and marketing (Van den Ban, 1998) to sustain themselves in agriculture profession. Increasing numbers of small marginal farmers and narrowing ratio of field extension personnel and farmers demands different approaches in agricultural extension to make it demand driven.

Extension Approaches

Number of extension approaches have been in practice in India. The general extension approach assumes that farmers are ignorant about availability of technologies and communication can lead to adoption while, commodity specific approach believes in bringing all development functions of a particular commodity under one umbrella. The Training and Visit system of extension relies on capacity building of extension worker and administrative reform while project approach presumes that development of particular area or group can be done in a time bound manner. All such approaches consider farmers as passive recipients and are mostly supply driven. On contrary, participatory and farming systems approach trust on wisdom of farmers and their active participation in development process. It further believes that many of problems, either technological or socio-psychological, can be solved through group action. The involvement of all stakeholders of agricultural development viz. farmers' organizations, research and extension organizations, and different support organizations (credit, marketing, etc.) is key to success of participatory approach with major say of farmers' organization in planning and execution of extension programmes.

The Cases of Group Approach

There is enough evidence that farmers' organizations, input organizations or output ones in form of Self Help Groups (SHG), Farmer

Interest Groups (FIG), Common Interest Groups (CIG), Commodity Cooperatives, Farmer Producer Companies/Organizations, etc., can articulate necessities of member farmers, arrange different kinds of services for them and can act as pressure group for influencing policies of government in their own favour. It provides its members with opportunities to participate in planning and implementation of different development programmes as well as have access to better inputs, resources, credits, markets and extension services. The Farmer Producer Organizations (FPO) provide producers, especially resource poor ones (small and marginal farmers), opportunities to come at common platform to form an successful coalition to jointly address challenges of agriculture faced by them like, access to credit, inputs, technology, processing, markets, etc. The farmers' producer organizations and producer companies are very much beneficial in improving value chain of agricultural produce and thereby proved to be in getting good prices for their produce. However, success of producer companies depends on farmers' commitment to company, integrity and quality of leadership, its acceptance within community, as well as market environment (Sawairam 2015).

The success of dairy cooperatives in India is well known which made India World's largest milk producer with 155.5 million tones milk production during 2015-16 that too with great contribution of small marginal farmers and landless labours. Besides, other cooperatives namely, Indian Farmers Fertilizer Cooperative, sugar cooperatives, etc., provide farmers, especially small and marginal ones, a mechanism for pooling their little resources for solving common problems relating to credit, supplies of inputs and marketing of agricultural produce.

Farmers of Uttarakhand, engaged in traditional beekeeping at subsistence level, were organized in groups by an NGO "Appropriate Technology India" and were introduced scope of developing a honey value chain involving improved wall hives, better swarm catching, safe extraction methods, seasonal bee management and an induced bee breeding technology. In collaboration with business service providers involved in construction and supply of bee boxes, swarm supply, maintenance of hives and certain beekeeping equipments, farmers' group increased its production base of more than 4400 bee colonies with annual production of 17.6 tonnes of honey. The tribal farmers of Thane and Nandurbar districts of Maharashtra organized themselves in CIGs and SHGs to transform their subsistence agriculture into sustained remunerative farming of flower cultivation and backyard poultry production, respectively through technical support and improved marketing channel from Bharatiya Agro Industries Foundation. To optimize income from small scale cultivation of pigeon pea (tur) in Western Vidarbha of Maharashtra, tur growers formed FIGs for processing of raw dal and its marketing. The FIGs set up mini dal mills with financial support from Sir Ratan Tata Trust increased income of its members through value addition in tur. The collective commercial production of perennial grass on van panchayat land by members of Women's dairy producer group in a village of Uttarkashi district of Uttarakhand provided an additional income to dairy producers due to increased milk production. The production of fodder has had a good impact in terms of increased milk production, because of which dairy producers have been able to generate a sustainable income" (Shrma *et al.*, 2013). These are examples of how collective work and group approach can solve many of problems and sustain production and income.

Based on research study conducted on Organic Producers Group of Uttarakhand and Indian Organic Producers Company Limited of Kerela, Cherukuri and Reddy (2014) concluded that leadership and its management was important factors in success. Leaders needed to constantly innovate themselves, being prepared not only for strengthening institution but also for meeting emerging challenges that were likely to crop up in view of market competitiveness.

Role of Krishi Vigyan Kendras

Krishi Vigyan Kendras (KVKs) under guidance of ICAR-Agricultural Technology Application Research Institutes have been instrumental in growth and development of existing FPOs, formed by different organizations, by providing technological back-stopping to their members. KVKs of Kerala and Tamil Nadu provided technological back-stopping in form of conducting skill training programmes to members and office bearers of FPOs of their respective districts on various facets of activities of FPOs (Murlidharan *et al.*, 2016, Sriram *et al.*, 20116). Besides, KVKs have also been contributing in formation, mobilization and evolution of FPOs. KVK Kannur facilitated Government of Kerala in implementing its poverty alleviation project "Kudumbashree" by forming 322 groups of goat farmers through organizing general orientation and skill training

(Giggin 2016). Likewise, KVK Erode prepared Farmers' Calendar depicting various aspects of FPOs for its promotion (Alagesan *et al.*, 2016).

Summary

The small holders in country are facing various social, economic, political and technological challenges. Despite these challenges and poor conditions of small and marginal farmers, productivity and profitability of their farms can be improved by its diversification towards high value low volume crops/enterprises through technological advancements and innovative institutional arrangements. The farmers groups provide producers, especially resource poor ones (small and marginal farmers), opportunities to come at common platform to form an successful coalition to jointly address challenges of agriculture faced by them like, access to credit, inputs, technology, processing, markets, etc., to improve value chain of agricultural produce and thereby getting good prices for their produce. Many KVKs are working with farmers groups and FPOs by providing desired technological back-stopping to their members. However, success of groups depends upon integrity and quality of leadership, its acceptance within community, as well as market environment. National Bank for Agriculture and Rural Development and other agencies are facilitating formation of farmers' groups and State like Maharashtra has come forward to operate large number of farmers' companies/FPOs which needs to be emphasized upon by other State Governments also. Small farmers' coming together may empower them to enhance volume of produce, ensure its quality, distribution of responsibilities within group and better marketability of produce.

Reference

Alagesan P., Maya Krishnan K. and Tamilselvan A. (2016). Status of Farmer Producer Organizations in Erode district.*Farmer Producer Organizations in Kerala and Tamil Nadu: A KVK Perspective*. ICAR-ATARI, Begaluru. Pp. 48-57.

Cherukuri, R. R. and A. Amarender Reddy. (2014). Producer Organizations in Indian Agriculture: Their Role in Improving Services and Intermediation.*South Asia Research*. 34 (3): 1–16.

Davis, K. Burton Swanson, David Amudavi, Daniel Ayalew Mekonnen, Aaron Flohrs, Jens Riese, Chloe Lamb, and Elias Zerfu. (2010). In-depth assessment of public agricultural extension system of Ethiopia and recommendations for improvement. *Discussion Paper 01041*, IFPRI, Washington, DC.

Giggin, T. 2 (2016). Farmer Producer Organizations in Kannur District, *In* Chandre Gowda M. J. and Sreenath Dixit (eds.) *Farmer Producer Organizations in Kerala and Tamil Nadu: A KVK Perspective*. ICAR-ATARI, Begaluru. Pp. 33-36

GoI.(2016).State of Indian Agriculture 2015-16, Government of India, Ministry of Agriculture & Farmers Welfare Department of Agriculture, Cooperation & Farmers Welfare, Directorate of Economics and Statistics, New Delhi

MahendraDev, S. (2013). Rising food prices in South Asia: A policy framework to mitigate adverse effects, Development Papers 1301, United Nations ESCAP, South and South-West Asia Office, New Delhi

Mukherjee, A., and Maity, A.(2015). Public–private partnership for convergence of extension services in Indian agriculture.*Current Science*. 109 (9 & 10):1557-1563.

Muralidharan P., Rajeev, M. S. and Ravi, S. (2016). Status of Farmer Producer Organizations in Alappuzha district.*Farmer Producer Organizations in Kerala and Tamil Nadu: A KVK Perspective*. ICAR-ATARI, Begaluru. Pp. 37-38.

Pal, S. (2017). Strengthening Delivery of Agricultural Extension Services in India: Experiences and Contemporary Issues, *Agricultural R&D Policy in India*. ICAR-NIAP. Pp. 33-52.

Sawairam, P. (2015). Case Study of Farmer Producer Organization in Maharashtra in Era of Globalization.*IBMRD's Journal of Management and Research*. 4 (2): 1-11.

Sharma P., Taneja S., Rawat P. and Chakraborty S. (2013). Krishi Sutra 2: Success stories of Farmer Producer Organizations. Department of Agriculture and Cooperation, Govt. of India.

Sriram N., Sudha A., Geetha P., Vijayakumar M., Srividya S., Kohila P., and Geetha B. (2016). Status of Farmer Producer Organizations in Salem district, *Farmer Producer Organizations in Kerala and Tamil Nadu: A KVK Perspective*. ICAR-ATARI, Begaluru. Pp. 39-47.

Van den Ban, A. (1998). Supporting farmers' decision-making process by agricultural extension. *Journal of Extension Systems*. 14:55–67.

□□□

Chapter-11

Millets for sustainable production system

M.G. CHANDRAKANTH[1] AND A. SEETHARAM[2]

[1]Director, Institute for Social and Economic Change, Bengaluru, [2]Former National Coordinator, AICRP on small millets and Professor Emeritus, ICAR, Bengaluru

The green revolution due to the continued support from research, extension and policy to rice and wheat, rendered the cultivators of small millets (fox tail millet, little millet, proso millet, kodo millet, barnyard millet), to shift to rice-wheat crops. Correspondingly millets were rendered inferior goods as their income elasticity of demand became negative and price elasticity of demand positive. Still, India continues to be the largest cultivator and consumer of millets with a negative growth rate in area and production (Table 1). Rice and wheat constitute 78 percent of the total food grains production and millets constitute 15 percent. Green revolution, by paving way for rice and wheat, reduced area under nutritive, healthy, cost effective and low water using millets, which were being cultivated by marginal and small farmers on marginal lands.

Table 1. Compound growth rates of area, production and productivity of major crops in India

(as per cent per annum with base TE 1981-82=100)

Crop	1980-81 to 1989-90			2000-01 to 2009-10		
	Area	Production	Yield	Area	Production	Yield
Rice	0.41	3.62	3.19	-0.03	1.59	1.61
Wheat	0.46	3.57	3.10	1.21	1.89	0.68
Jowar	-0.99	0.28	1.29	-3.19	-0.07	3.23
Bajra	-1.05	0.03	1.09	-0.42	1.68	2.11
Maize	-0.20	1.89	2.09	2.98	5.27	2.23
Ragi	-1.23	-0.10	1.14	-3.03	-1.52	1.57
Small Millets	-4.32	-3.23	1.14	-5.28	-3.58	1.78
Barley	-6.03	-3.48	2.72	-1.41	-0.25	1.17
Total Coarse Cereals	-1.34	0.40	1.62	-0.76	2.46	3.97
Total Cereals	-0.26	3.03	2.90	0.09	1.88	3.19
Gram	-1.41	-0.81	0.61	4.34	5.89	1.48
Tur	2.30	2.87	0.56	0.26	1.82	1.56
Other Pulses	0.02	3.05	3.03	-0.34	-0.32	0.02
Total Pulses	-0.09	1.52	1.61	1.17	2.61	1.64
Total Foodgrains	-0.23	2.85	2.74	0.29	1.96	2.94
Sugarcane	1.44	2.70	1.24	0.77	0.93	0.16
Oilseeds	1.51	5.20	2.43	2.26	4.82	3.79
Cotton	-1.25	2.80	4.10	2.13	13.58	11.22

Source: http://indiabudget.nic.in/es2010-11/echap-08.pdf

Even though the MSP for millets are almost on par or even higher than rice and wheat, farmers are not responding to their cultivation due to:

1. Lack of procurement operations
2. Lack of consumer demand due to niche market, and
3. Poor value addition in millets.

The rice and wheat are procured at huge transaction cost. For instance, while Government procures rice at ₹ 1000 per quintal, it spends ₹ 1043 per quintal towards distribution, and wheat at ₹ 1120, spending ₹ 424 per quintal towards distribution. The transaction cost of rice procurement is 104 percent of rice price, while that for wheat is 38 percent of wheat price.

Value addition in millets can reverse Engel's law

Value addition and awareness due to technology can perhaps reverse the income and price elasticities of demand (as in the case of maize). Maize, a coarse grain which was an inferior good turned out to be a normal and superior good due to value addition not only as poultry feed but also as corn flakes, corn flour, corn oil, corn starch and so on. Similarly, Ragi (*Eleusine coracana*) (a small millet) is turning out to be a superior good gradually due to value addition, awareness and capacity building explaining the medicinal properties notably its richness in calcium. Thus, value addition in millets can convert them from inferior to normal and superior goods, apart from fetching socio economic, environmental and natural resource benefits withwelfare implications (Fig. 1).

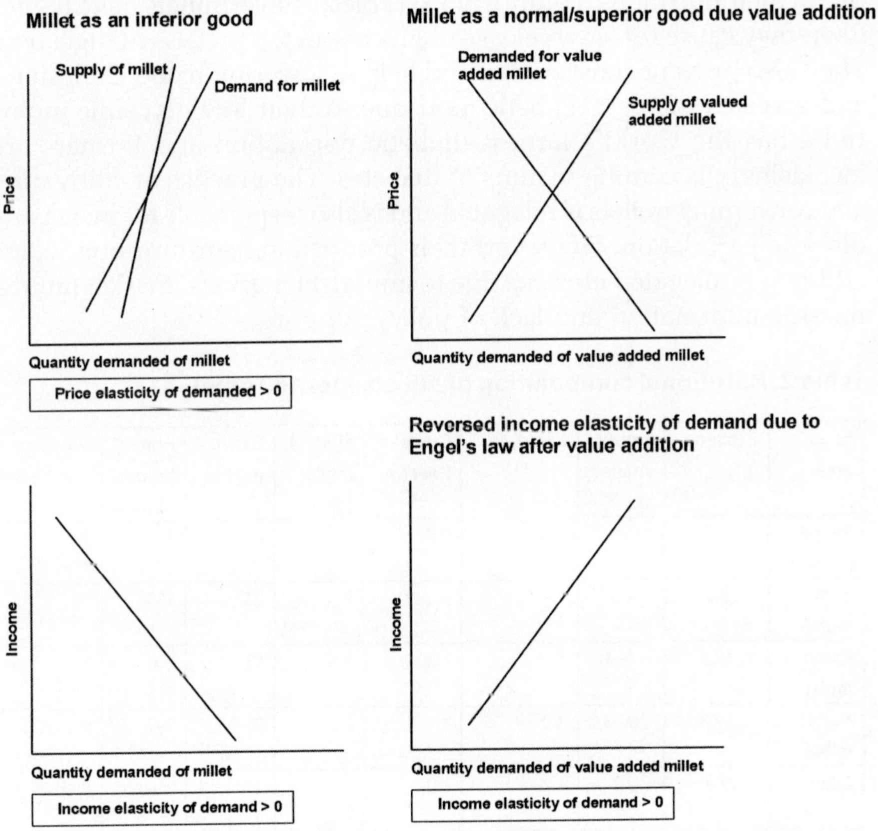

Fig. 1. How inferior goods can be normal/superior goods after value addition

Threat to food security

India's food security is threatened by climate change. Rain-fed agriculture contributes 40 percent of India's food production, where small and marginal farmers who constitute 80 percent of India's farmers are predominant. In addition, they are resource poor, capital poor, with little or no access to markets and have barely limited livelihood opportunities. Millets are grown on 35.46 million ha in the world producing 28.52 million tons of which, India shares 60% of area and 65% of production being the largest cultivator, producer and consumer of millets. The productivity is 8.76 quintals per ha due to lack of input use and low market prices due to lack of demand as a result of support to major cereals. Among the grains, small millets, foxtail millet and proso millet outperform cereals such as wheat and rice in their nutritional composition (Table 2). Since millets have higher fiber, they cause for slow release of glucose in the process of digestion. They also prevent constipation and help in lowering blood cholesterol and serve as the best diabetic food due to their low glycemic index. India has the world's largest diabetic population and farmers are increasingly becoming victims of diabetes. The practice of cultivating and consuming millets is relegated and is also responsible for increasing diabetic population. However, their production, consumption, value addition is relegated *inter alia* due to imperfect markets, market failure, missing information and lack of policy support.

Table 2. Nutritional composition of millets (per 100 gms)

Food grain	Protein (g)	Carbohydrates (g)	Fat (g)	Crude fiber (g)	Mineral matter (g)	Calcium (mg)	Phosphorous (mg)	Iron (mg)
Finger millet	7.3	72	1.3	3.6	2.7	344	283	3.9
Kodo millet	8.3	65	1.4	9	2.6	27	188	12
Proso millet	12.5	70.4	3.1	7.2	1.9	14	206	10
Foxtail millet	12.3	60.9	4.3	8	3.3	31	290	5
Little millet	7.7	67	4.7	7.6	1.5	17	220	6
Barnyard Millet	6.2	65.5	2.2	9.8	4.4	11	280	15
Wheat	11.8	71.2	1.5	1.2	1.5	41	306	5.3
Rice	6.8	78.2	0.5	0.2	0.6	45	160	

Source: All India Coordinated Small Millets Improvement Project, A Profile, Indian Council of Agricultural Research, University of Agricultural Sciences, Bangalore (undated)

Green Revolution Largely Due to Irrigation

According to IWMI currently, over 80% of irrigated agriculture in India is supported by groundwater, resulting in severe overexploitation of this resource[6]. As green revolution was largely for wheat and rice which are water intensive, its growth is unsustainable unless India switches towards small millets. Small millets are drought tolerant and thrive in harsh agroclimatic conditions. Hence they indirectly conserve the precious and expensive groundwater resource. The small millets and gender are closely associated as women who have preserved the culture of both cultivating and cooking millets also form the major family labor. A majority of the farm operations are carried out by women in these areas. Cultivation of millets not only is low external input demanding, but also reduces physical stress, and also complements nutritional needs of the farm family. However, the post harvest processing of millets is in infancy and with no policy support to millets, millets are further relegated. Thus, applied field and lab oriented research on nutrimillets would address the key issues of natural resource conservation, food, nutritional, livelihood and economic security, gender balance in the wake of climate change.

Millets serve the dual purpose of meeting food, nutrition as well as fodder security. The socioeconomic, cultural aspects of cultivation of climate smart small millets such as fox tail millet, kodo millet, proso millet, little millet, pearl millet, finger millet and Barn yard millet, is that they not only require low water but also have nutrition levels much higher than rice or wheat. Hence their impact on household food, nutrition, health, livelihood and economic security including economics of cultivation of millets for different classes and locations of farmers with environmental sustainability are crucial.

The millets such as Little, Proso, Foxtail, Kodo are the most nutritious but their popularity faded resulting in negative growth in area, production and consumption, as they cannot compete with other remunerative crops due to poor market, policy support and consumer awareness. The extension efforts towards their cultivation and consumption are relegated. Advent of irrigation in the villages gradually replaced the area under millets with rice, maize and commercial crops.

---

[6] http://southasia.iwmi.org/Data/Sites/15/Documents/PDF/IWMI_South_Asia_Brochure,2005)

Due to low market price their cultivation is uneconomical, but there is market failure as the market fails to capture the nutritive value of millets. Millet cooking is thus popular in dry lands, however the recipes may have undergone changes in irrigated areas, peri urban and urban areas. Nowadays the demand for millets is increasing especially due to increasing diabetic population and hence may become popular due to rise in demand.

Advantages of Shifting to Millets

1. Provides the best agronomic climate smart practices in terms of conservation of scarce water resources by way of improved water use efficiency leading to food, nutrition, fodder and heath security
2. In semi arid tropics water is economically and physically scarce. Small millets offer the best economic solution to accommodate food, fodder, nutrition and health security for the benefit of small and marginal farmers
3. The existing processing facilities can be used to procure and process small millets towards primary value addition.

The benefits for climate change adaptation and mitigation are:

1. Improved household food, fodder, nutrition and health security through cultivation of climate smart millets
2. Improved water use efficiency in economically and physically scarce water conditions
3. Climate change resulting in increased temperature over years is wiping out water and groundwater and hence millets are the only solution in such areas which are increasing in India

The small millets foxtail, kodo, proso, barn yard, little, pearl and finger millets are replicable in all semi arid and arid regions of the world which are fraught with acute water and groundwater scarcity. There is a need to unlearn the current practice of heavy focus on rice-wheat ecosystem in order to accommodate these crops, which were previously grown in areas currently occupied by superior cereals.

Cultivation of Small Millets on Rainfed / Dry Lands

The millets such as Little, Proso, Foxtail, Kodo are the most nutritious but their popularity faded resulting in negative growth in area, production and consumption, as they cannot compete with other remunerative crops due to poor market, policy support and consumer awareness. The extension efforts towards their cultivation and consumption are relegated. Advent of irrigation in the villages gradually replaced the area under millets by rice, maize and commercial crops. Due to low market price their cultivation is uneconomical, but there is market failure as the market fails to capture the nutritive value of millets. Millet cooking is thus popular in dry lands, however the recipes may have undergone changes in irrigated areas, peri urban and urban areas.

Little Millet JK8

Fox Tail Millet - Krishnadevaraya

Kodo Millet GPUK 3

Proso Millet TNAU 145

Barn Yard Millet VL 172

Little Millet OLM 203

The PL 480 Grains

Green revolution did bring food security to India putting a halt to starvation deaths and 'ship to mouth' existence of the PL480 project era. However, in the post green revolution period, India continued to emphasize on the superior cereals (rice and wheat) and despite the richness in nutritional and medicinal properties of millets and emphasis on organic agriculture, small millet cultivators (growing fox tail millet, little millet, proso millet, kodo millet) are relegated and these crops are gradually disappearing in traditional areas. Small millets are now inferior goods with negative income elasticity of demand and positive price elasticity of demand.

The post harvest processing, value addition, awareness, capacity building of farmers, consumers, processors, and market linkages can bridge the constraints both producers and consumers are facing in respect of meeting the millet needs. Paradoxically while producers are unable to find the market, consumers are unable to access millets even though they are in need, especially the diabetic population. It is hence necessary to bridge this gap.

Knowledge base, Complements, Low marketable surplus, Status symbol

Knowledge regarding cookingmillet rich food is with (elderly) farm women belonging to different castes. The preparations are traditional/ ethnic and specific to certain castes in the villages. While cooking millet dish, the complements are crucial and are in vernacular (Eg. Ragi rotti - Hucchallu Chutney; ragi huriittu with milk, ghee and jaggery). However, as eating habits die hard, small millets are cultivated on a minute area in the farm, to meet immediate farm family requirements and hence with low marketable surplus. Most often farmers who cultivate millets are also consumers of those millets. Three is also a status symbol attached to consumption of millets (and indirectly the cultivation of millets) as their appearance may not be as fine/neat/great and palatable compared with rice/wheat.

Subsistence scale cultivation, Low external input users

Cultivation of millets is limited to farm level and not market oriented as the width of the market is getting narrowed and effective demand is falling. Data on farmers cultivating them/cooking them are not documented. Snowball sampling is crucial to locate them. Crop duration of millets is smaller compared to cereals and at least 10 to 15 percent of the crop duration is saved with the cultivation of millets (for Eg. Fox tail millet comes to harvest in 75 days). They are users of low external inputs as it is not so economical to use agrochemicals and they are also tolerant to pests and diseases.

What is Required for Green Development

For sustainable agricultural growth it is necessary to promote food, nutritional and economic security through sustainable cultivation of nutri-millets by small and marginal farmers in semi arid regions of India, with value addition in the wake of climate change. This requires baseline study of the small millet cultivators covering income, employment, savings, investment, gender and natural resource dependence and coping mechanisms in the wake of climate change. The socio-economic, cultural aspects of cultivation of small millets and their impact on household food, nutrition, health, livelihood and economic security including economics of cultivation of millets for different classes and locations of farmers with environmental sustainability are crucial. The techno-economic assessment of existing post harvest technologies and for value addition, for applied research and innovations to standardize the post harvest technology for nutrimillets is necessary. Provision of backward and forward linkages for value chain in nutrimillets using the standardized post harvest technologies through domestic and international markets is desirable. Capacity building for social capital formation among farmers and consumers regarding their cooperation for millet cultivation and consumption considering their nutritional, medicinal properties is the need of the hour. The support from Government through extending procurement support is more than sufficient as the market is picking up for millets. Currently the price of millets is around ₹ 10,000 per quintal at the retail level. However there is market imperfection since producers are not getting even 20 percent of the consumer's price. This calls for removal of market imperfections, notifying the commodity in APMC, and procuring the millets from farmers. For the current market trends which are towards millet consumption to make an impact on producers, it is crucial to reduce market imperfections and see that the farmers receive a fair proportion of consumer rupee. This can happen through market extension, innovations in technologies of millet processing and procuring millets from farmers.

□□□

Chapter-12

Sustainable development through small farm horticulture production system

BALAKRISHNA, B.[1], ATHEEQULLA, G.A.[2] AND AIYASHA, G.F.[3]

[1] Principal Scientist, [2] Scientist and [3] Senior Research Fellow, ICAR-Indian Institute of Horticultural Research, Bengaluru

Indian agriculture is dominated by small farm production systems, where availability of land and other resources is highly limited. Per capita availability of land is declining (2.30 ha in 1970-71 and 1.15 hectares in 2010-11), which is a direct indication of increasing proportion of small and marginal farmers (The proportion of total area operated by marginal farmers increased from nine percent in 1970-71 to nearly 15 percent in 1990-91). This situation is demanding

i) Development, refinement and promotion of suitable technological innovations to different agro-ecological zones and also to micro farming situations,
ii) Specific crops and production situations for extensive and focused institutional support and
iii) Development and expansion of efficient group/cooperative and sustainable marketing strategies to completely utilize quality produce of small farm production systems in agriculture in India.

Among various sub-sectors of agriculture contribution of horticulture is gradually increasing over a period of time with current production of 283.6 million tonnes in 2015-16, surpassing total agriculture production of country (251.57 million tonnes in 2015-16). The increased production trend is mainly due to multiple reasons like

1. Increase in area under different horticultural crops through introduction of many promotional schemes of both central and state governments.

2. Increased expectations and demand for innovations from farmers, including quality planting materials, in different horticultural crops like fruits, vegetables, flowers, medicinal and aromatic crops, tuber crops, spices, plantation crops, mushrooms, etc.
3. Large scale introduction of innovations in multiple horticultural crops (new and improved high yielding varieties, scientific nursery management for production of good quality seeds and planting materials, integrated production packages, protected cultivation, etc.) by both public and private institutions, which have significantly contributed for increase in productivity and quality of horticulture produce along with improved post harvest management and marketing to cater needs of both domestic and export markets.
4. Use of multiple and integrated extension methods by ICAR, SAUs and other private institutions to cater specific technological requirements of individual and group of horticultural farmers, which is directly helping farmers to adopt innovations quickly.

Majority of horticultural crops are cultivated in small production systems, where in efficiency of resource utilization, higher productivity, quality and net-profit are prime concern of farmers. These factors are essential to make horticultural enterprises resource efficient, sustainable and profitable, without which farmers cannot continue farming. There is a large number of horticultural activities in management of horticultural crops like selection and production of quality planting materials of new varieties / hybrids in different crops; integrated water, nutrient, pest and disease management; production and application of bio-pesticides; botanical pesticides and bio-fertilizers; organic farming; protected cultivation with different types of structures and good post harvest and marketing management for domestic and export markets. When activities and practices are adopted holistically, farmers and other stakeholders can achieve higher productivity, quality along with efficient management of resources. To synchronize promotion of holistic packages for horticultural crops with market demand, multi-disciplinary team of scientists of ICAR-IIHR is extending their knowledge with their rich experience of two decades in using Integrated Technology Dissemination System (ITDS) among small farm horticultural production systems. This action research experience was not only encouraging for researchers learning but also making farmers highly successful in pursuing their objective of increasing yield, productivity, quality and net- profit in horticultural crops and enterprises. The results of implementation of holistic extension approach in some programmes like NABARD, NICRA and RKVY projects are given in Table 1, 2, 3 and 4. In all these projects, small and marginal farmers constituted > 80% of total participants of project.

Table 1. Impact of demonstration* of climate resilient technologies (CRT'S) on yield of horticultural crops (N=33) under NICRA - TDC project of IIHR

S. No.	Crops	Average yield (in tonnes/acre/ annum)			Gross income (Rs.) per Acre	Net income (Rs.) per Acre
		Before CRT	After CRT	% increase		
1.	Tomato	21.80	32.8	50.45	262400	1,98,179
2.	Watermelon	09.77	16.6	69.90	160000	1,10,934
3.	Cucumber	23.50	32.25	37.23	161250	1,18,750
4.	Capsicum	14.27	21.90	53.46	444166	3,92,041
5.	Field Beans	10.40	16.80	61.53	168000	1,19,545
6.	Onion	12.00	15.80	31.67	126400	83,949
7.	Bitter guard	12.60	17.50	38.88	140000	84,476
8.	Papaya	28.00	42.50	51.78	340000	2,77,036

* Data collected from demonstration farmers' fields under NICRA project during 2012-13 and 2013-14

Impact of demonstration of climate resilient technologies, as a holistic package, is clearly seen in enhancing yield in different vegetable crops before and after adoption of climate resilient technologies (CRT's) ranging from 32% to 70% depending on crop. Gross and net-income was also in range of ₹ 1.26 lakhs and ₹ 0.84 lakhs per acre in a 75 days short duration crop like cucumber to ₹ 4.44 lakhs and ₹ 3.92 lakhs per acre in 5½ months crop like capsicum. It was a sustained technical support and marketing information that has resulted in such increased yield and income of farmers.

There are problems like incidence of pests and diseases, lack of quality seeds, cold chain storage facilities, and sanitary and phyto-sanitary measures, etc. These problems are hindering growth of both production and export of vegetables. Popularization of high-yielding cultivars, cold chain transport facility, and other infrastructure facilities would be required for improving performance of both production and export of vegetables (Sidhu, A.S., 2002). Keeping in view such production difficulties, ICAR-IIHR integrated package was customized to give holistic and improved package to address production and other field based problems to promote vegetable production under both open cultivation and protected environment. This has resulted in higher yields with better quality produce suitable for both domestic and export markets.

Table 2. Yield (t/ac) levels of different vegetables among PCV farmers* (N= 60)

Category of farmer	Coloured Capsicum	European Cucumber	Green Capsicum	Tomato
Marginal Farmers	42.50	34.10	53.80	62.40
Small Farmers	38.60	28.50	48.24	58.30
Big Farmers	36.20	24.45	46.40	55.20
Average	39.10	29.01	49.48	58.63

* Data collected fromFarmers of RKVY project on "Demonstration of Biological Control of Pests and Diseases and Integrated Crop Management Practices in Protected Cultivation of Vegetables and Flowers" implemented by ICAR-IIHR, Bengaluru.

In India, in present scenario of perpetual demand of vegetables and shrinking land holdings drastically, protected cultivation is best alternative and drudgery-less approach for using land and other resources more efficiently (Sirohi and Bahera, 2000). To keep up requirements of time and demands of industry, ICAR-IIHR spread holistic package for intensive and timely management packages for few commercially viable crops. Due to complete adoption of ICAR-IIHR packages, yield levels of four important commercially cultivated vegetable crops under protected environment were high among all small and marginal farmers, highest being in tomato (62.40 t/ac) followed by green capsicum (53.80 t/ac) among marginal farmers, when compared to big farmers, as given in Table 2.

It was very clearly observed from data in table 3 that a large majority of farmers (86.87%) were small & marginal land holding farmers engaged both in open and protected cultivation of vegetable crops. Among protected cultivation farmers also >70 % constitute small&marginal farmers, who prefer poly-house structure over other types of protected structure due to many advantages such as :

i) Safety and durability of structure and quality and stability in production

ii) Good protection from rain & other erratic climatic changes and

iii) Two and a half to three times higher yield with assured quality of produce with better crop management.

Thus, small and marginal farmers are more consciously produce good quality in addition to obtaining higher yields. Nagarajan *et al.*, (2002) also opined that Greenhouse is most practical method of accomplishing objectives of protected cultivation.

Table 3. Categorization of farmers based on their land holding and nature (open/ protected) cultivation of vegetable crops

S. No.	Production environment	No. of farmers in different Categories (Based on land holding)			
		Marginal (< 1 ha)	Small (1 to 2 ha)	Big (> 2 ha)	Total
1.	Open cultivation	38	64	13	115
2.	Protected cultivation	13	24	08	45
	i. Polyhouse	10	18	06	34
	ii. Net house	3	6	2	11
	Total (1 + 2)	**51**	**88**	**21**	**160**

When technologies were disseminated with holistic approach and market integration, acceptance and adoption of technologies were very high and resulted in expected results in terms of higher yield and income to farmers, reduction in use of chemicals and pesticides, increased use of bio-pesticides and botanical insecticides, enhanced application of soil resilient products and finally quality and safe (farmers, consumers and environment) production of horticultural crops. Complete participation of all stakeholders at different stages of production and marketing would also increase larger benefits to farming community, more so for small and marginal farmers, who work under resource constrained conditions.

Demand-driven extension is desirable in many instances, although a balance must be struck between demands of government and those of farmers in different economic categories (e.g. estate, emerging, low-income, and marginal). Striking a balance between institutional supply systems and farmer-initiated demand-driven extension/technology systems should in many cases be ultimate goal of countries eager to advance to higher stages of development and competitive power (Chikaire *et al.*, 2012).

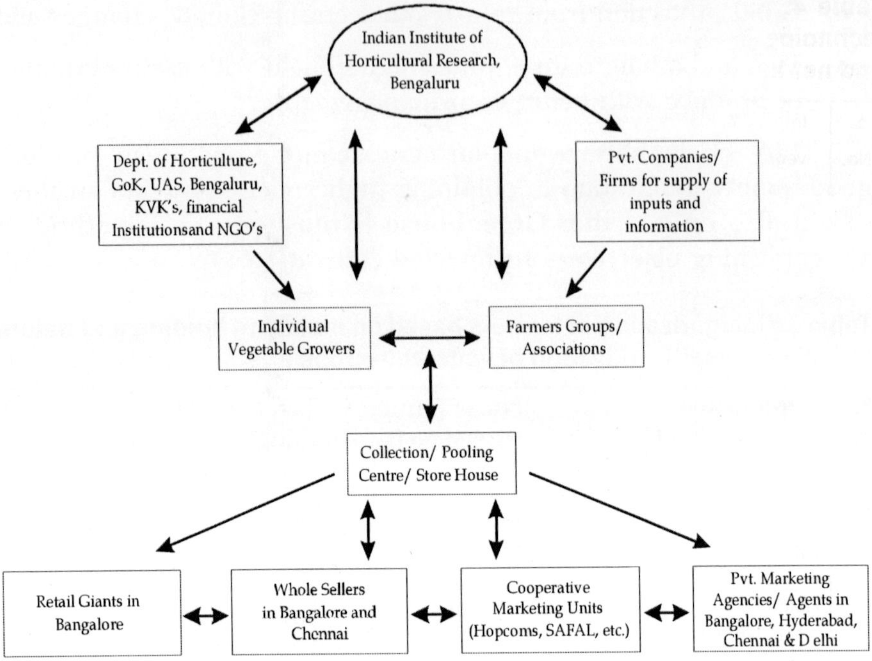

Fig. 1. A system approach for Promotion of safe and quality vegetable production through Market led Technology Transfer system

Sustainability in production and marketing of horticultural produce can be achieved with integration of innovations and institutions, which has been successfully demonstrated under various projects and programmes of ICAR-IIHR, more so in recently implemented NABARD and RKVY projects to promote protected cultivation. Dissemination of holistic package for protected cultivation crops was planned and implemented with market driven technologies. An example of this technology dissemination module was and its impact on yield and quality of produce along with percent produce sent to export is given in Table 5 and Figure 1. This ITDS module is very useful not only demonstration of technologies but also very effective in continuance of adoption of these holistic market driven packages for a very long period (so far 8 years). This method of promotion of agricultural innovations builds and sustains technology adoption and creates sustainable development of selected production system, more so among small and marginal farmers.

Table 4. Impact of dissemination of integrated vegetable production technologies (ITDS) on change in investment on crop management, gross and net-income of vegetable growers

S. No.	Important vegetable crops cultivated by farmers	Investment for crop production (Rupees per acre)		Gross Income (Rupees per acre)		Additional increase in net-income (Rs. per acre) from use of ITDS of IIHR
		Before ITDS	After ITDS	Before ITDS	After ITDS	
1.	Tomato	48,500.00	68,800.00	1,28,000.00	2,14,000.00	86,000.00
2.	Capsicum	67,000.00	1,08,000.00	1,36,000.00	2,68,000.00	1,32,000.00
3.	Okra	53,000.00	86,500.00	86,000.00	1,44,000.00	58,000.00
4.	Pole beans	48,600.00	63,500.00	69,000.00	1,38,000.00	69,000.00
5.	Field beans	48,600.00	53,500.00	62,300.00	1,24,000.00	61700.00
6.	Water melon	24,800.00	49,500.00	58,000.00	1,42,000.00	84,000.00
7.	Cucumber	42,250.00	73,500.00	82,700.00	1,61,250.00	78,550.00
8.	Onion	34,600.00	42,400.00	51,500.00	93,800.00	42,300.00
9.	Cabbage/ Cauliflower	38,500.00	53,500.00	62,300.00	1,08,000.00	45,700.00
10.	Bitter Gourd	58,600.00	66,300.00	76,200.00	1,40,000.00	63,800.00
11.	Polyhouse cultivation of Coloured Capsicum	4,35,000.00	5,30,800.00	8,32,000.00	12,56,400.00	4,24,400.00
12.	Net-house cultivation of Coloured Capsicum	3,85,000.00	4,64,800.00	5,62,000.00	8,26,400.00	2,64,400.00
13.	Net-house cultivation of European cucumber	1,65,000.00	2,04,800.00	2,72,000.00	3,26,400.00	54,400.00

* Data collected from beneficiary farmers of RKVY projects of ICAR-IIHR, implemented during 2013-14

There is a visible impact of use of ITDS in spreading of horticultural technologies among farmers of small horticultural production systems, in terms of increased yield and net-income. Even though farmers have made little more investment for adoption of technological innovations, in long run it yields larger benefits, i.e., increase in net-income ranging from 18% to 50% in different crops.

In all these projects and programmes implemented by multi disciplinary team of scientists of ICAR- IIHR adoption of innovations is not only complete, highly effective and successful but also created sustainable socio-economic impact. Therefore, integration of technological innovations and their promotion through holistic extension approach (use of multiple methods of extension) along with strengthening and participation of multiple institutions creates sustainable ways of learning and adoption of innovations by farmers; creates market integration with pre-production tie-ups; farmers consciously plan for total quality management among their group/ cluster of producers and creates good infrastructure for crop production, development of storage of fresh and processed produce, better transportation and sustained marketing which is need of hour for sustainable production of small farm horticulture production systems in India. These integrated and holistic approaches not only make persons, institutions, resources and technologies more productive but also create stronger sustainable development of horticulture industry in country.

Hence, effort of promoting innovations should be on a sustainable basis which can be achieved through

1. Integration of technologies or innovations and use of multiple methods of extension
2. Strengthening of institutional linkages, effective utilization of resources of multi- institutions
3. Creating ambience for sustained learning and adoption of innovations by farmers
4. Providing market integration for quality produce of farmers
5. Creating good infrastructure for crop production, development of storage of fresh and processed produce, better transportation and sustained marketing.

When all these facilitating factors are available, like one developed and disseminated by ICAR-IIHR programmes, it is possible to create sustainable production among small farm horticulture production systems in India. These approaches not only make persons, institutions, resources and technologies more productive but also make entire process highly sustainable.

Tomato produced by NICRA adopted village farmers ready for dispach

Good and healthy onion crop grown in rainfed condition with CRT's of IIHR

Polyhouse grown coloured capsicum crop ready for harvest in a small farmer field

Grading and packing of coloured capsicum grown in polyhouse

A view of tomato crop in polyhouse

Reference

Balakrishna, B., Hegde, M.R. and Srinivas, V. (2008). Market led extension approach for promotion of quality and safe exotic commercial vegetable production and marketing: A field experience. Golden Jubilee and National Seminar on Innovative Extension Strategies for Agricultural Development and Rural Prosperity from Dec 18-20, 2008 at Department of extension Education, RAU, Pusa. PP:168.

Balakrishna, B.*et.al.* (2014). *Climate Resilient Technologies (CRT's for Rainfed Onion Growers to Enhance Yield, Quality and Income.* Published by Director, IIHR, Bengaluru. 2014.

Chikaire, J. *et al.* (2012).*Continental J. Sustainable Development.* 3(1): 1-18.

Nagarajan, M., Senthilvel, S. and Palanysamy, D. (2002). Material substitution in Green house construction. *Kisan World.* 11: 57-58.

Sidhu, A.S. (2002). Current status of vegetable research in India. www. tropical-seed.com

Sirohi, P.S. and Bahera, T.K. (2000). Protected cultivation and seed production in vegetables. *Indian Horticulture.* 45: 23-25.

❑❑❑

Chapter-13

Scaling up milk production and sustainability in smallholder production system: Avenues and approaches

K P RAMESHA[1] AND A KUMARESAN[2]

[1] Head, ICAR-NDRI, SRS, Bengaluru

[2] Principal Scientist, ICAR-NDRI, SRS, Bengaluru

Livestock is an important sub-sector of Indian agriculture, providing livelihood to the population and also a source of food, income and employment. As per the BAHS, 2012, in India there are about 56.7% of world's buffaloes (105.05 million) with 1st position, 12.5% cattle (199.07 million) with 2nd position, 20.4% small ruminants comprising 144 million goats and 72 million sheep with 2nd and 3rd position, respectively in the world. Livestock sector is highly dynamic and contributes substantially to food and nutritional security to masses. It is evolving in response to surging demand for livestock produce due to population growth, sustained income and economic growth, urbanisation, burgeoning middle income class and changing lifestyles. India ranks first among the world's milk producing Nations since 1998 with 160.35 MT in 2015-16.

The different production systems like zero input - low output, low input - moderate output, intensive input - high output are characteristics of dairy production system in India. Smallholder farming comprising of landless, small and marginal farmers is very common in India, of which a majority of households rear dairy animals either as a mainstay and/or complementary to crop production. The contribution of dairy animals to livelihoods of millions of people of this segment is well beyond the production for household consumption or for market. Besides their monitory benefit and providing a steady stream of food and revenues for households, dairy animals provide employment to the family, act as insurance during crop failures, contribute to gender equality by generating opportunities for women, generates *in situ* fertilizers for enhancing the soil fertility, contributes to day-to-day expenses of the farm family, recycle waste products and residues from cropping or agro-industries and supplies energy source for cooking. Although mechanization is taking place at greater speed, still a major proportion of farmers in the region depend upon the draught power of bullocks for farming. Thus any progress of dairy sector results in balanced development of the rural economy, particularly in reducing the poverty amongst the weaker sections, as the ownership of the livestock is more egalitarian compared to land and other productive assets. The rural women play a significant role in dairy husbandry and are directly involved in most of the operations relating to feeding, breeding, management and health-care of the animals. However, in spite of the significant contribution of the smallholder dairy sector in socio-economic up-liftment of the resource poor farmers and the overall agricultural growth, full potential of this sector has remains untapped.

Contribution of Smallholders to the National Milk Pool

Currently the global milk demand is growing by 15 million tons per year, mostly in developing countries. This increased volume of milk is being produced by small-scale dairy farmers, and millions of jobs per year may be created in primary production. This presents a unique opportunity for building up a sustainable dairy chain that sources milk from smallholder dairy farmers to meet not only the demands of local consumers but also those of the world market. While capitalizing on this opportunity could generate significant wealth in rural areas and provide benefits to all stakeholders involved in the dairy value chain, it calls for a sound dairy development strategy. Since smallholder dairying is considerably affected by factors such as resource access, service delivery, food safety standards as well as national and international subsidies, effective strategies are to be evolved considering all these factors. If the technological competence of the rural people in dairying is substantially improved, it would not only improve the self-employment and also enhance the rural economy and sustainable livelihood.

The major factors affecting the milk production in smallholder system includes, but not limited to :

- Poor Productivity of animals and inadequate availability of superior germplasm
- Inadequate feed and fodder supply
- Problems in diffusion of new technologies
- Mind set of people to manage animals under low/negligible input system
- Indiscriminate breeding of animals at field conditions
- Reduced availability of vaccines and other health measures
- Unorganized marketing

How have we reached first position in milk production?

Till date, smallholder dairying is the backbone of dairying in the country and contributes to more than 70 percent of the total milk produced. It is not a simple task to increase the milk production from 17 million tonnes in 1950-51 to 160.3 million tonnes during 2015-16. Dairy farmers, scientists, co-operative agencies and development departments have all played a pivotal role to reach this level of milk production in the country. The huge population of dairy animals is not the only reason for increasing milk productivity. Analysis of milk production based on "animal population based increase" and "productivity based increase" indicate that more than 50% of enhancement in milk production came from the productivity based increase. Thus it is imperative that the scientific advancements favored technology mediated increase in milk production in the country.

We started crossbreeding of indigenous breeds with exotic high yielding breeds in 1960s and currently the contribution of crossbred cattle to total milk production is around 24%. The Indigenous cattle and buffaloes contribute to 21 and 51%, respectively. Although the crossbreeding contributed significantly to improve milk production, it also resulted in dilution of our Indigenous breeds of cattle. Further, the adaptation of temperate exotic breeds or their offspring out of crossing with Indian breeds need sophisticated and scientific management to exploit their full production potential under tropical climate of India. Yet another important problem is the disposal of crossbred males; since they are less preferred for draft purpose and for agricultural operations, a majority of the farmers do not wish to rear crossbred males and as such we do not have a well-defined mechanism for their disposal. All these have become a cause of concern about crossbred cattle.

What did it cost us?

Genetic erosion of indigenous breeds is a problem of national concern and a number of local breeds are at the risk of extinction. As on date, we have registered 40 breeds of cattle and 13 breeds of buffaloes in the country. The unique feature of these animals is that they were developed under varying agro-ecology, climate and production systems. Since time immemorial, these animals provide milk, work power and organic fertilizers to agriculture and employment to rural mass besides acting as a means for food and nutritional security. The existing indigenous breeds have been evolved over the centuries through natural selection for adoption to harsh and poor management conditions in different ecological niches. Indigenous breeds of cattle are well adapted to harsh climate, resistant to common tropical diseases and can subsist on poor quality crop residual roughages, grasses and scanty drinking water. They are also well known for heat tolerance, hardiness and ability to survive and perform even under stressful conditions.They produce moderately under low input system. Maintenance and management of this valuable vast diversity has become a challenge as most of these breeds are low producers, facing genetic dilution or even complete erosion due to many factors like increasing pace of organization, mechanization of agriculture, over emphasis on some high producing breeds and many unforeseen factors.

Some indigenous breeds are having good potential of milk production (Gir, Sahiwal, Rathi, Red Sindhi, Thaprparkar, etc). Such breeds are mostly evolved in North-West part of the country. The lactation milk yield of these breeds varies between 1200-3000 litres or even more. Such breeds needs to be promoted and may also be used for grading up of non-descript cattle in different parts of the country for improving their milk productivity. Due to its unique characteristics like hardiness under unfavorable climatic conditions, breeds like Sahiwaland Gir were imported from India by wide list of countries and regions. Using Sahiwal, two Australian tropical dairy breeds, the

Australian Milking Zebu and the Australian Friesian Sahiwal have been developed by Australia. The Gir breed is introduced in Brazil in 1911 and gained popularity from 1940s. Recent claims indicate that Gir cow now records over 62 litres/day in Brazil. While Indian cattle breeds are doing exceptionally fine abroad, their productivity has not been fully exploited in India. Indian cows and buffaloes have a rich A2 allele gene which helps them produce healthier milk. The frequency of this A2 allele in Indian breeds is 100 per cent whereas in exotic cattle breeds it is less than 60 per cent. Imported breeds posses A1 allele, which is considered to be associated with diabetes, obesity and cardiovascular diseases. The need of the hour is to have concerted efforts to convert the strength of having huge indigenous population into reality in terms of high milk production.

I. Improving the Availability of Superior Animals

Since the availability of good quality indigenous animals is very much limited, the conservation, improvement and multiplication of the elite animals assumes much significance and need to be taken up with at most priority. In this regard, with the existing set up, Gaushalascan as focal point for development and conservation of Indigenous breeds. For the protection of cows, a total of 4187gaushalas have been established of which 1413 is registered by AH department as on 2014. The government is assisting several gaushalas by providing financial grant. Some of the gaushalas are having good infra-structure and capable of acting as conservation cum genetic improvement units. There is a need to develop a strong linkage between research institutions working on conservation aspects with these gaushalas.Gaushalas can play four important roles, besides several others, in development and conservation of indigenous breeds.

1. **Genetic improvement centers:** Genetic improvement is a time taking process and requires vast network of agencies and dedicated man power with great desire for the cause. Since there are several dedicated gaushalas doing great in this regard, first we need to identify the good gaushalas having good number of animals belonging to specific indigenous breeds. These identified gaushalas can act as focal point for genetic improvement of identified breed(s). The first expectation out of these gaushalas could be production of quality breeding bulls. As we all know, there is a huge demand for quality semen; currently we produce about 96 million frozen semen doses and we need around 140

million doses to cover at least 50% of the breedable population in the country. The identified gaushalas should spare out at least 10% quality elite females (milk production above the breed average) for production of quality bulls. These gaushalas can be linked with identified semen station in that state. The male calves produced out of these females should be given premium management practices and they should be taken in by the semen stations for making them good bulls and for production and supply of quality semen doses to field conditions for breed improvement. Through this process we can reduce the demand and availability of semen doses and also achieve genetic improvement.

2. **Technology demonstration centers:** Identified gaushalas can be developed into technology demonstration centres. The promising production, feeding, health and reproduction techniques can be applied first at the identified gaushalas and it can act as a focal point for method or result demonstration of the technologies. Farmers in that area can be taken to the gaushala frequently for first hand exposure. For greater benefit, gaushalas in a particular region can be linked up with KVKs in the region. Because most of the KVKs do not have a big dairy animal farm, it will be a big help to the KVK personnel for demonstrating their proven technologies to the farmers in association with the gaushalas.

3. **Dairy Training school:** Since a good number of animals are available with gaushalas they can act as a regional training centres for developing trained manpower in the area of dairying. The farmers starting small scale dairying can obtain first hand training in dairy animal management practices.

4. **Organic dairy production model:** Identified gaushalas that rear animals in a better way without much external inputs and antibiotics can be gradually converted into organic dairy production models. If enough land is available for cultivation of crops and fodder, even they can be converted into organic mixed farming units. It can be used to show the farming community that "Waste is not a waste but a misplaced resource". This facility should contain all the requirements for recycling most of the by-products of one unit into input for other units. This may contain units for vermicompost, panchkavya, vermiwash, integrated dairy-fish culture models, units for processing cow urine etc so that the model is viable and self-sustainable. Provisions are to be provided to link the gaushala to market also.

II. Integrated crop-livestock farming for generation of in situ inputs

Traditionally our livestock production has essentially been a mixed production system involving few heads of ruminants and poultry along with other components like agriculture, horticulture and fisheries. This integrated mode of production offered scope for *in situ* generation of inputs for different components besides offering security to the farmers in case when one system fails due to unforeseen natural calamities. However, during recent years we have witnessed a gradual transformation from "integrated" to "intensive" livestock production owing to increasing demand for livestock products. Although the intensive single-component oriented systems has been shown to increase production, the sustainability of this system-presently as well as-in future depends upon several issues. For instance, intensive mode of livestock production essentially requires quality feed and fodder in desired quantity. However the existing situation and the forecasted estimates are not encouraging. In 2010, the demand for green and dry fodder was 1061 and 589 million tones however the availability was only 395 and 451 million tones leading to a deficiency of 63 and 23%, respectively and in future the gap between demand and availability is expected to increase. While the population of livestock is increasing, the area under fodder producing crops remained by and large at constant. Further, on account of diversified useof agriculture residues, the gap between thedemand and supply of fodder is increasing. The utilization pattern of Maize, which constitutes a major part of livestock feed, indicates that 52 and 11% of the maize produced in the country were utilized as poultry and cattle feed, respectively however till date we are not able to meet the requirement of livestock industry. The situation is further aggravated by the current policy giving emphasison use of grains for bio-fuel production. Thus, the grains like maize,rapeseed and soybean are likely to be diverted towardsproduction of bio-fuels, which will reduce the availability of these grains for livestock feeding.

Decreasing soil fertility is another concern which would affect the feed and fodder supply to the livestock. The intensive input agriculture, although helped us to improve the unit land productivity, with rampant use of chemical fertilizers, pesticides, herbicides, fungicides and insecticides compromised the soil fertility. Due to imbalanced nutrient application, non-judicious use of pesticides, intensive cropping system, decline in soil biodiversity and depletion of organic matter in soil sustaining the land productivity is the biggest question before us. Increased fuel prices and transportation charges are other concerns, which would increase the cost of livestock feeding. The entire feed and fodder transport system is based on the availability of cheap fuel, which is no longer a guarantee. Given this situation what is the solution to making available the required feed and fodder for livestock now and in future? Recalling our age old livestock production would give an indication to integrate several components along with livestock so that a majority of the inputs are generated *in situ*. When crop-livestock production system was in place, the animals helped to sustain the soil fertility and crop yields through their manure. During earlier days about half of the total dungproduced by the livestock was used as manure to fertilize the crop fields. For instance, in 1970-71 dungmanure accounted for about 43% of the total value of manure and fertilizersused in agriculture, which declined drastically to 23% in 1980-81 and to about13% during the 1990s. During these days several commercial livestock farms experience problems in disposal of farm wastes including manures. If a farm has its own grain and fodder production area or bio-gas production unit, these manures and wastes can be recycled as a valuable input for fertilizing the crop land.

III. Community animal management, milking and processing facilities

The biggest task in front of the dairy industry is to transform, gradually and phase by phase, the unorganized dairy sector into a more profitable organized dairying which insures the farmer's profit and also safeguards the consumer's interest both, at price and quality level. The usual concept of Corporate Large Scale Dairying is considered to be having a negative impact on the unorganized traditional dairy production system. Hence, the concept of Community Animal Management System (CAMS) can be conceptualized for the overall benefit of the smallholder farmers and also to fulfill the increasing market demand with due consideration of the quality thereby bringing smallholders into the organized umbrella.

The idea is to be conceptualized after making a ground level survey both for the production system of the locality and the socio-economic status of the farmers. The major constraint in adopting modern animal rearing practices is the small holding of animal unit. Locality having a milking cattle and buffalo population of 400-500 is to be identified as a single Module. The module will have a single shed for all the animals of the locality with all necessary facilities for modern scientific dairy management. All the animals will be equipped with proper identification number and to be insured. The modern feeding systems will be practiced as per the requirement e.g. total mixed ration, region-wise mineral mixture feeding, balanced ration etc. Properly trained subject-matter specialists may be appointed for routine management practices. Automated Milking System may be installed with sensors and wireless network system, which otherwise is never possible in small scale dairying. All the required milk production parameters may be transmitted to a single server where it can be stored and retrieved to work as a Database System. The database will be of immense help for Field Data Recording System and the sensors for various milk parameters can be used for early diagnosis of certain on-coming diseases like mastitis and many other metabolic diseases. As there is minimal involvement of human labour which otherwise can be diverted for more productive work, more revenue can be generated and also better the quality of product. The waste (cow dung and other waste materials) may be utilized for biogas production and slurry to be used as manure for the crop cultivation. The profit may be distributed as per the total fat corrected milk produced (after standardization) from all the animals of an individual farmer.

Summary

Smallholder farmers have been the back bone of Indian dairying. Any dairy development measure taken up to improve the milk production should keep them in mind before implementing any policy. The major problem with smallholder production system is lower milk productivity and meeting the stringent quality standards of modern dairy chains is the biggest challenge in front of the smallholders as their production system, at least partly, does not follow the "Clean Milk Production" practices. The 'Good Agricultural Practices' have to be applied that not only include aspects related to quality and food safety, but also aspects of economic, social, and environmental sustainability. The most successful AMUL model needs to be replicated in several untouched areas for successful participation of smallholders in the modern dairy supply chains. For increasing the production of value added dairy products, the infrastructure will need to be further developed at the public, cooperative as well as at private sector level. The Institutes and Universities engaged in dairy education and research need to enhance their efforts to provide research support and human resources for large scale, diversified and quality production of value-added products. The traditional unorganized marketing of milk in smallholder system needs to be gradually shifted towards organized marketing for better and sustainable remuneration to the producers.

□□□

Chapter-14

Climate-smart smallholder dairy production system

MUKUND A KATAKTALWARE[1], NAZAR, S[2], LETHA DEVI G[3] AND KP RAMESHA[4]

[1]Senior Scientist, [2]SRF and [4]Head ICAR- NDRI, SRS, Bengaluru
[3]Senior Scientist, ICAR-NIANP, Bengaluru

Since time immemorial animal husbandry is an integral part of India's agricultural economy and performs diversified role in providing livelihood security to the rural populace. As per the 19th Livestock Census, 2012, India has the largest livestock population in the world i.e. 512.05 million including 300 million bovines. According to estimates of the Central Statistics Office, in 2015-16, the value of output livestock sector at current prices was about ₹ 5,91,691 which is about 28.5% of the value of output from agricultural and allied sector (Anonymous, 2017). In India, dairy sector has become an important secondary source of livelihood for millions of rural families. Dairying plays an important role in providing employment and income generating avenues for small, marginal farmers and landless labourers in general and women in particular. Smallholder dairy production system is the characteristic feature of Indian dairy sector. In 2015, the world average dairy farm size was three cows (comprised of 90 countries representing about 98% of milk production worldwide), whereas the average Indian dairy herd consists of two animals (USA-182 cows, New Zealand-419 cows). Most of the milk in the country is produced by the smallholders i.e. small, marginal farmers and landless labourers. Several measures initiated by the Government resulted in increasing the milk production significantly from the level of 17.0 million tonnes in 1950-51 to 155.5 million tonnes in 2015-16 and thus making India largest producer in the world. In the past, rising demand for milk was attributed to rapid population growth, however, now increase in per capita consumption has become an additional growth driver. With ever increasing human population and economically strong potential consumers, the demand for milk and milk products is likely to increase rapidly.

However, there are several challenges that are likely to obstruct the growth of dairy sector in India. Apart from the existing challenges like increased occurrence of emerging and re-emerging animal diseases, vulnerability to exotic diseases, perennial shortage of feed and fodder resources and an urgent need to increase production as well as productivity to meet ever increasing demand for animal products, etc., now dairy sector is confronting a very serious challenge of 'climate change' (Kataktalware *et al.,* 2016). The probability of extreme high temperature events (heat waves) would increase as mean temperature increases, and an increasing number of heat waves could significantly increase the negative impact of global warming, especially on livestock (Klinedinst *et al.,* 1993). The variability in climate viz., a decrease in minimum temperature, an increase in maximum temperature, erratic rainfall, higher incidence of extreme weather events, etc., are likely to affect immune response, productive and reproductive efficiency, and welfare of dairy animals (Kataktalware *et al.,* 2017), which in long term would affect the livelihood and food security of millions of farmers' dependent of livestock. Such volatile variability poses an alarming challenge to biological production systems, which may affect the sustainability of smallholder dairy production system. Even though the change in climate is a global phenomenon, its adverse effects are more severely experienced by poor people in developing countries who rely heavily on the natural resource base for their livelihoods (IFAD, 2009).

Climate Change and its Impact on Indian Dairy Sector

Change in climate is a natural phenomenon, however, after the industrial revolution, the rate of change has increased, and in last couple of decades, it is found to be alarming. Now it widely accepted that the human-caused climate change is real, and GHG emissions are causing changes to the planet that could possibly trigger dangerous consequences by the turn of the century. Climate change could severely change the distribution and quality of natural resources which may adversely affect the livelihood security of millions of people worldwide. The vulnerability of the agriculture in general and dairy sector in particular to climate change is well documented. The vulnerability is likely to be acute in light of technological, resource, and institutional constraints (Ayyappan and Shivanna, 2017). Although the prospect of animal productivity and its share to food security and national GDP is on upswing, livestock is adversely affected by the prevailing scenario of heat stress, land and water scarcity and the ensuing climate change (Srivastava and Singh, 2017). India is categorized among the most vulnerable countries because rural economy is dependent on crop-livestock production systems, and almost 70 per cent of livestock in India is owned by smallholders. Animals of such milk producers with inadequate resources are more susceptible to impacts climate change, and thus are at greater risk.

Climate change may have a substantial effect on water availability that could cripple the dairy economy like never before. Direct effects of environmental stress on livestock include temperature-related sickness and death, and the morbidity of animals. Indirectly it influences emergence of pathogens, distribution of vector-borne diseases, host immunity, scarcity of feed and water, etc. which affects animals' health. The magnitude of loss will be particularly high in the states that have coastal areas, as the Temperature Humidity Index (THI, an indicator of heat stress on animals); the load is higher in the coastal areas due to high relative humidity (Sirohi and Sirohi, 2010). The annual milk of buffaloes is likely to come down by more than 100 litres in Haryana, Rajasthan, and West Bengal. This study estimated that the milk production at the all-India level would decrease by about 7.85 MT, even if the number of lactating dairy animals remains equal to their population in 2003. The increase in mean THI by 2 units could further aggravate the production losses to 9.65 million tonnes, about 23% more than the losses incurred by 1 unit increase in mean THI. In value terms aggregate losses under alternate scenarios of THI increase amount to whopping 140.65 billion and 172.84 billion (2007-08 prices), that is 8.68 and 10.66 percent, respectively of the current (2007-08) value of the output of milk and milk products. India is currently losing nearly two percent of the total milk production, amounting to whopping over 2,661 crore due to rise in heat stress among cattle and buffaloes because of the global warming (Upadhyay *et al.*, 2007). The negative impact of direct temperature rise on total milk production of cattle and buffaloes for India has been estimated about 1.6 MT in 2020 and more than 15 MT in 2050 (Upadhyaya, 2014). The potential losses in milk production in 2030 from elevated heat stress due to decline in productivity of dairy animals as estimated by Sirohi and Sirohi (2010) is presented in Table 1.

Table 1. Annual loss in milk production in 2030 from elevated heat stress

States	Mean annual THI increase = 1unit				Mean annual THI increase = 2unit			
	Crossbred cows	Local cows	Buffaloes	Total annual loss (million liters)	Crossbred cows	Local cows	Buffaloes	Total annual loss (million liters)
	(liters/animal/year)				(liters/animal/year)			
Andhra Pradesh	697.0	15.0	52.9	433.0	856.5	18.4	65.0	532.0
Assam	640.0	20.8	73.5	109.9	786.6	25.6	90.3	135.1
Bihar	716.3	27.6	97.3	514.6	880.2	33.9	119.6	632.4
Chhattisgarh	56 7.7	10.0	35.4	38.4	697.6	12.3	43.5	47.2
Gujarat	740.2	26.7	94.4	467.5	909.6	32.8	116.0	574.4
Haryana	665.0	36.0	127.3	339.9	817.2	44.3	156.4	410.3
Jharkhand	745.5	25.6	90.3	95.6	916.0	31.4	110.9	117.4
Karnataka	319.0	1.3	4.5	169.6	391.9	1.6	5.6	208.4
Kerala	838.2	10.2	36.0	499.2	1030.0	12.5	44.3	613.5
Madhya Pradesh	618.9	16.2	57.1	283.5	760.5	19.9	70.2	348.3
Maharashtra	532.4	8.3	29.3	421.1	654.2	10.2	36.0	517.4
Orissa	817.1	27.3	96.5	271.8	1004.1	33.6	118.6	334.0
Punjab	610.0	18.2	64.1	488.3	749 .6	22.3	78.8	600.0
Rajasthan	659.6	39.6	139.7	808.3	810.6	48.6	171.7	993.3
Tamil Nadu	888.4	20.4	72.1	1475.3	1091.6	25.1	88.6	1812.8
Uttar Pradesh	616.5	20.8	73.6	979.5	757.5	25.6	90.4	1203.6
Uttrakhand	224.3	4.1	14.4	19.3	275.6	5.0	17.7	23.7
West Ben gal	828.7	35.6	125.7	443.3	1018.3	43.7	154.5	544.8
All India				7852.0				9648.6

Adaptation to climate change

Adapting appropriate strategies would ensure sustainability of smallholder production system under changing climate scenario. An advanced planning of production management systems is required, with an understanding of animal responses to thermal stress and ability to provide management options to prevent or mitigate adverse consequences (Nienaber and Hahn, 2007). The environmental stress caused due climate variability could be ameliorated by selection of thermo-tolerant animals, microclimate modification, nutritional manipulation, preventive health care, etc.

1. **Animal selection**: Our country is bestowed with 40 indigenous breeds of cattle and 13 breeds of buffalo. The milch breeds of cattle like Sahiwal, Tharparkar, Gir and Red Sindhi, dual purpose breeds of cattle like Rathi, Kankrej, Ongole, Deoni, etc. and buffalo breeds like Murrah, Jafarabadi, Bhadawari, Surati etc., are well adapted and known to have unique characteristics like better heat tolerance, diseases resistance and milk constituents. Such well-adapted breeds should be preferred over exotic/ graded crossbreds of Holstein Friesians (HF). As compared to HF crossbreds, Jersey crossbreds have fewer maintenance requirements, better heat tolerance, higher disease resistance, better milk quality with comparable production capacity. Another alternative approach could be to change the species, e.g. small ruminants like goat instead of cattle or buffalo. Instead of giving emphasis on milk production of just one or two lactations alone, the dairy farmers should take into account the lifetime productivity of animals, prevailing geo-climatic conditions and availability of resources and technical knowhow while choosing the species and breed of dairy animals. The long-term sustainability of production system should be the goal rather than short-term gains from high yielding exotic breeds alien to local climatic conditions.

2. **Microclimate modification**: The first step to reducing heat stress is to make provision of cool water and shade for all animals. Under extensive farming system, provision of community shelters with potable water should be made so that animals can take rest during peak hot hours. Reduction in total heat load to the extent of 30 to 50% can be achieved with a well-designed shade. Trees having large canopy can be planted in paddocks as it not only protects from direct sunlight and imposes bare minimum radiant heat load but also cools the surrounding air without interfering with air circulation. The animal shelters should have an east-west orientation as it reduces the amount of direct sunlight shining on the side walls or entering the house. Thatch, earthen tiles, corrugated iron sheet, an iron sheet with aluminium painting, etc. could be used as roofing material. Shade cloths like woven polypropylene which are less expensive could also be considered as a roofing option as it provides about 80 to 85% as much shading. Outer coverings and shading, perhaps combined with a roof-spray, are popular greenhouse technologies that can

be applied in the structural design of dairy animal housing to reduce the height of the eaves. Various roof systems give improved natural ventilation by means of roof openings, enhanced solar chimney effect, etc. A well-ventilated animal housing facility is necessary for providing fresh, clean air to animals and workers. Based on the number of animals, fans should be selected to provide appropriate winter, mild weather, and summer ventilation rates. Placing fans in tactical locations throughout a dairy facility is the first step in providing supplemental cooling for the cows, assuming that adequate natural ventilation already exists and plenty of fresh water is always available. Additional cooling may be necessary for lactating dairy cows in hot, humid environmental conditions. Air movement (fans), wetting the cow, evaporation to cool the air, and shade to minimise transfer of solar radiation can be used to enhance heat dissipation.

3. **Feeding strategies**: Under stressful environmental conditions provision of and easy access to cool water, changed feeding schedule (at least 60 percent during cooling hours i.e. 8 pm to 8 am), increasing the energy density of the diet, using high quality forages and feeding more concentrates should help the dairy animals maintain its energy requirements even though consumption of dry matter is reduced. During lean season, other alternative strategies to be adopted are feeding hay, silage, Azolla, tree and vegetable leaves, compressed complete feed block, hydroponics fodder production, etc. Further, supplementation of vitamin-A, β-Carotene, E, B-complex vitamins, selenium, copper, zinc, magnesium, potassium along with buffers viz., sodium bicarbonate, sodium sesqui-carbonate, magnesium oxide, etc. may be considered as it would help to reduce thermal stress and maintain feed intake. The smallholder dairy farmers should be aware of the modern agronomic practices for the production and conservation of fodder, which would improve the supply of animal feed reducing undernourishment.

4. **Health care**: During environmental stress, the immune status of dairy animals is compromised. The maintenance of hygiene and sanitation in farm premises, proper housing with adequate floor space per animal, balanced feeding, prophylactic vaccination, prompt treatment and regular screening of animals against major contagious diseases would help to reduce the impact of thermal stress on animal health.

5. **Other strategies:**Other climate-smart practices which need to popularised among dairy farmers are:
   - Cultivation of drought-resistant fodder shrubs
   - Utilization of unconventional feed resources
   - Conservation of surplus fodder for lean season
   - Development of community fodder banks
   - Rainwater harvesting to meet summer requirements
   - Restoring and rejuvenating lakes to meet livestock requirements
   - Increasing water productivity by optimal utilization of water
   - Maximum use of recycled water for routine operations
   - Utilizing non-renewable energy resources for running farm operations
   - Integrating feasible technologies with prevailing production system

Mitigation of GHG emissions

The major sources of Green House Gas (GHG) emissions in the dairy production system include the enteric fermentation of animals, manure disposal, and production of feed and forage. Methane is one of the important green house gases and its production in India is estimated to be 105 Tg/year, out of which about 60 per cent is produced by enteric fermentation of ruminants, 92 per cent of which is produced only by cattle and buffalo and eight per cent is generated by all other herbivorous animals (Kamra, 2017). The animals grazing the lowest quality pasture will produce the highest amount of methane per unit product (Kirschgessner et al., 1995). The treatment of paddy with straw with urea resulted in a significant improvement in digestibility of nutrients and a significant depression in methane generation by the animals (Sahoo et al., 2000). Methane emission from animals may be less on feeding leguminous forages than feeding of grasses (Malik *et al.*, 2013). Malik and Singhal (2008) documented a significant reduction in methane production on the inclusion of limited quantity of lucerne fodder (30%) to the wheat straw based ration. Plant secondary metabolites such as tannin, saponin, essential oil, etc. showed promising results under in vitro conditions and need to be evaluated further for their action in animal systems by looking at the nutrient availability and productivity of animals (Malik *et al.*, 2013). Improving animal waste management like covered storage facilities, bio-gas production, electricity generation, etc., would reduce GHGs like methane and nitrous oxide significantly.

Summary

Smallholder dairy farmers are the backbone of Indian dairy sector. The climate change is likely to impact performance of dairy animals, which affect sustainability of smallholder dairy production system. The existing adaptation and mitigation strategies have potential to significantly reduce likely impact of climate change on dairy production system. It is high time that intensive efforts are made to augment the resilience of the farms and animals through dissemination of adaptation and mitigation strategies so that the production and productivity levels of dairy animals are not only maintained but also improved even in hostile climatic conditions. The timely adoption of climate-smart practices would definitely contribute in evolving sustainable a smallholder dairy production system.

Reference

Anonymous. (2017). Annual Report 2016-17. Department of Animal Husbandry Dairying and Fisheries, Government of India.

Ayyappan, S. and Shivanna, H. (2017). Foreword. In: Agriculture under climate change: Threats, strategies and policies. Belavadi *et al.*, (Editors), Allied Publishers Pvt. Ltd. Bengaluru.

IFAD. (2009). Livestock and climate change. Retrieved from: https://www.ifad.org/documents/10180/48b0cd7b-f70d-4f55-b0c0-5a19fa3e5f38.

Kamra, D.N. (2017). Production of methane by the livestock and its mitigation techniques. In: Agriculture under climate change: Threats, strategies and policies. Belavadi *et al.*, (Editors), Allied Publishers Pvt. Ltd. Bengaluru, Pp: 261-267.

Kataktalware, M.A., Nazar, S., Letha Devi G. and Ramesha, K.P. (2016). Adaptation and mitigation strategies for sustainable dairy production under changing climate scenario. In: Climate Change Challenge (3C) Socio-Economic-Ecological Interface-Building Exploring Potential Adaptation Strategies for Bio-resource

Conservation and Livelihood Development. S. Nautiyal *et al.* (Editors). Springer International Publishing Switzerland 2016. Pp: 519-530. DOI 10.1007/978-3-319-31014-5_31.

Kataktalware, M.A., Letha Devi G. and Ramesha, K.P. (2017). Climate smart dairy production. XIII Agricultural Science Congress- 2017 on Climate Smart Agriculture organized by University of Agricultural Sciences, Bengaluru and NAAS, New Delhi at UAS, Bengaluru during February 21-24, 2017. Pp: 217.

Kirchgessner, M., Windisch, W. And Muller, H.L. (1995). Nutritional factors affecting methane production by ruminants. In: Ruminant Physiology: Digestion, Metabolism, Growth and Reproduction. Aelhardt *et al.* (Eds). Ferdinand Enke Verlag, Stuttgart. Pp: 333-348.

Klinedinst, P.L., Wilhite, D.A., Hahn, G.L. and Hubbard, K.G. (1993). The potential effects of climate change on summer season dairy cattle milk production and reproduction. *Climate Change.* 23:21–36.

Malik, P.K. and Singhal, K.K. (2008). Influence of lucerne fodder supplementation on enteric methane emission in crossbred calves. *Indian Journal of Animal Sciences.* 78: 293–297.

Malik, P. K., Bhatta, R. and Prasad, C. S. (2013). Nutritional and biological interventions for enteric methane mitigation. In *Management strategies for sustainable livestock production against impending climate changes,* Kataktalware, *et al.* (Eds.) Model Training Course, Southern Regional Station, ICAR-NDRI, Bengaluru, India.Pp: 54-60.

Nienaber, J. A. and Hahn, G. L. (2007). Livestock production system management responses to thermal challenges. *International Journal of Biometeorology.* 52: 140-157.

Sahoo, B., Sarawat, M.L., Haque, N. And Khan, M.Y. (2000). Energy balance and methane production in sheep fed chemically treated wheat straw. *Small Ruminant Research.* 35: 13-19.

Sirohi, S. and Sirohi, S.K. (2010). Vulnerability of milk producers to climate change: Technological and policy options for livelihood security. In Upadhyay, *et al.* (Eds.), *Climate Change & Livestock Productivity in India.* National Symposium, ICAR-NDRI, Karnal, India. Pp:143-152.

Srivastava A.K. and Singh, S. (2017). Climate resilient livestock farming. In: Agriculture under climate change: Threats, strategies and policies. Belavadi *et al.*, (Eds), Allied Publishers Pvt. Ltd. Bengaluru. Pp: 34-37.

Upadhyay, R.C. (2014). Physiological determinants of climate resilient animal production and future global challenges. In P.S. Yadav, et al. (Eds.), *Physiological determinants of climate resilient and sustainable animal production.* National Symposium, ICAR-CIRB, Hisar. Pp: 31-39.

Upadhyay, R.C., Singh, S.V., Gupta, A.K. and Ashutosh, S.K. (2007). Impact of climate change on milk production of Murrah buffaloes. *Italian Journal Animal Science.* 6: 1329-1332.

□□□

Chapter-15

Institutional innovations in context of climate change

TRIVENI DUTT

Joint Director (Academic) and Dean, ICAR-IVRI, Izatnagar

Climate change is any long-term significant variation in average weather of a given region. It involves changes in variability or average state of atmosphere over a duration which may range from decades to millions of years. These changes can be caused by dynamic process on Earth, external forces including variations in sunlight intensity, and more recently by human activities.

The Intergovernmental Panel on Climate Change (IPCC) concluded that most of observed increase in globally averaged temperatures since mid-twentieth century is likely due to observed increase in anthropogenic greenhouse gas concentrations via greenhouse effect. The greenhouse effect was discovered by Joseph Fourier in 1824 and was first investigated quantitatively by Svante Arrhenius in 1896. On Earth, major greenhouse gases are water vapor, which causes about 36–70% of greenhouse effect; carbon dioxide (CO_2), which causes 9-26%; methane (CH_4), which causes 4- 9%; and ozone, which causes 3–7%.

Methane is a more effective greenhouse gas than carbon dioxide, but its concentration is much smaller so that its total radioactive forcing is approximately four times less than that from carbon dioxide. Some other naturally occurring gases contribute very small fractions of greenhouse effect; one of these, nitrous oxide (N_2O), is increasing in concentration due to human activity such as agriculture. The three main GHGs are carbon dioxide (CO_2), methane (CH_4), and nitrous oxide (N_2O). While most attention has focused on carbon dioxide, methane and nitrous oxide (both extremely potent GHGs) have greater global warming potentials (GWPs) than CO_2. By assigning CO_2 a value of one GWP, warming potentials of these other gases are being expressed on a CO_2-equivalent basis. CH_4 has a GWP of 25 while N_2O has a GWP of 296.

Due to green house effect, surface temperature is increasing. The global average air temperature near Earth's surface rose by 0.74 ± 0.18 °C (1.33 ± 0.32 °F) during hundred years ending in 2005. (IPCC, 2007). Climate change is likely to have major impacts on livestock as heat stress potentially impair production and increase vulnerability to various diseases. Furthermore, climate change indirectly affects supply of feed and fodder and water resources.

Climate Change and Livestock Production

Physiologically in a thermal environment, in which animal's heat production exceeds heat loss, an increasing amount of heat is stored in animal's body, resulting in increased body temperature. When body temperature is significantly elevated, numerous of homeothermic events are initiated which are directed towards maintaining homeothermy. However, feed intake, metabolism, body weight and milk yield decrease to help alleviate heat imbalance. It is believed that deviation of any of functions beyond a certain threshold or critical temperature level by environment may modify growth lactation or reproduction processes. Not only production, growth, reproduction is affected by climate change but health is also compromised. The major impacts of climate change on livestock and human diseases have been on diseases that are vector-borne. Increasing temperatures likely to support expansion of vector populations into cooler areas as well as increase in number in warmer areas. Changes in rainfall pattern can also influence an expansion of vectors which may lead to more disease outbreaks. Climate changes could also influence distribution pattern of various disease indirectly through changes in distribution of livestock. For example, small ruminants sheep & goat, which are well adapted to arid region, and if precipitation will increase, mechanisms responsible for maintenance of homeothermy may not be sufficient to survive under climate change scenario. Under hot climate, livestock may become more vulnerable to diseases due to lowered immune status.

Climate Change and Smallholders Agriculture Production System

Livestock are responsible for 12% of anthropogenic greenhouse gas emissions. Sustainable intensification of livestock production systems might become a key climate mitigation technology. However, Indian livestock production systems vary substantially, making implementation of climate mitigation policies a formidable challenge. Indian agriculture is represented by small and marginal farmers. Therefore, future of sustainable agriculture growth and food security in India depends onperformance of small and marginal farmers. The 'smallholders agriculture' has been defined by Burnett (1997) as "farming and associated activities which together form a livelihood strategy where main output is consumed directly, where there are few if any purchased inputs and where only a minor proportion of output is marketed". The smallholders farmers will suffer significantly due to climate change, but it is hard to predict impact. The factors which will increase vulnerability of smallholder farmers include-Small farm size, low use of technology, low capitalization, and diverse nonclimate stressors etc. However, it is resilient too because of family labor, patterns of diversification away from agriculture and possession of indigenous knowledge.

Adaptation of Livestock Sector to Climate Change

IFAD in their thematic paper have identified ways to increase adaptation in livestock sector.

1. Production system adjustments

Changes in livestock practices could include:

i) Diversification and intensification of livestock and crop production.

ii) Changing land use and irrigation.

iii) Altering timing of operations.

iv) Conservation of nature and ecosystems.

v) Introducing mixed livestock farming systems, such as stall-fed systems and pasture grazing.

2. Breeding strategies

Many local breeds are already adapted to harsh living conditions. Livestock breeding programmes and agricultural programmes may suitably be modified that might help to speed adaptation. Adaptation strategies address not only tolerance of livestock to heat, but also their ability to survive, grow and reproduce in conditions of poor nutrition, parasites and diseases (Hoffmann, 2008). Such measures could include:

i) Identifying and strengthening local breeds that have adapted to local climatic stress and feed sources and

ii) Improving local genetics through cross-breeding with heat and disease-tolerant breeds. If climate change is faster than natural selection, risk to survival and adaptation of new breed is greater (Hoffmann, 2008).

3. Institutional and policy changes

Removing or introducing subsidies, insurance systems, income diversification practices and establishing livestock early warning systems and other forecasting and crisis-preparedness systems–could benefit adaptation efforts.

4. Science and technology development

Working towards a better understanding of impacts of climate change on livestock, developing new breeds and genetic types, improving animal health and enhancing water and soil management would support adaptation measures in long term.

5. Capacity building for livestock keepers

There is a need to improve capacity of livestock producers to understand and deal with climate change increasing their awareness of global changes. In addition, training in agroecological technologies and practices for production and conservation of fodder improves supply of animal feed and reduces malnutrition and mortality in herds.

6. Livestock management systems

Efficient and affordable adaptation practices need to be developed for rural poor who are unable to afford expensive adaptation technologies. These could include:

i) Provision of shade and water to reduce heat stress from increased temperature. Given current high energy prices, providing natural (low cost) shade instead of high cost air conditioning is more suitable for rural poor producers.

ii) Reduction of livestock numbers–a lower number of more productive animals leads to more efficient production and lower GHG emissions from livestock production (Batima, 2006).

iii) Changes in livestock/herd composition (selection of large animals rather than small).

iv) Improved management of water resources through introduction of simple techniques for localized irrigation (e.g. drip and sprinkler irrigation), accompanied by infrastructure to harvest and store rainwater, such as tanks connected to roofs of houses and small surface and underground dams (FAO, 2008; Thornton, *et al.*, 2008; Sidahmed, 2008).

Summary

Climate change is a major challenge for agriculture, food security and rural livelihoods for millions of people including poor in India. Adverse impact will be more on small holding farmers. Farmers those living in already fragile environments, face risk of crop failure, loss of livestock. Social study of impacts of climate change on poor rural people in developing countries has tended to be concerned with increased frequency of extreme events. Climate change would have adverse impact on food security and livelihoods of smallholder's farmers. Serious interdisciplinary efforts must be initiated to generate scientific knowledge of effects of climate change on crops and livestock to "complex, diverse and risk-prone" farming systems of developing countries. The coping strategies would be useful to have long term adaptation strategies.

Reference

Barnett, A. (1997). *AIDS Briefs: Subsistence Agriculture* (USAID Health and Human Resources Analysis and Research for Africa Project).

Batima, P. (2007). *Climate change vulnerability and adaptation in livestock sector of Mongolia*. Assessments of impacts and adaptations to climate change. International START Secretariat, Washington DC, US.

FAO.(2008). *Climate-related Transboundary Pests and Diseases Including Relevant Aquatic Species*. Expert meeting, FAO. February, 2008.

Hoffmann, I. (2008). *Livestock Genetic Diversity and Climate Change Adaptation*. Livestock and Global Change conference proceeding. May 2008, Tunisia.

IPCC. (2007). *Climate Change 2007: Summary for Policymakers of Synthesis Report of IPCC.*Fourth Assessment Report. Draft copy, 16 November.

Livestock and climate change–Livestock Thematic Papers: Tools for project design. (Available at https://www.ifad.org/documents/10180/48b0cd7b-f70d-4f55-b0c0-5a19fa3e5f38)

Sidahmed, A. (2008). *Livestock and Climate Change: Coping and Risk Management Strategies for a Sustainable Future.* In Livestock and Global Climate Change conference proceeding, May 2008, Tunisia.

Thornton P., Herrero M., Freeman A., Mwai O., Rege E., Jones P. and McDermott J.(2008). *Vulnerability, Climate Change and Livestock – Research Opportunities and Challenges for Poverty Alleviation.* ILRI, Kenya.

❑❑❑

Chapter-16

Institutional innovations to cope with climate change in agricultural sector

LALITH ACHOTH[1], RESHMA M.[2] AND K.B.VEDAMURTHY[3]

[1] Professor and Head, [2] Professor and [3]Assistant Professor
Dairy Science College, KVAFSU, Hebbal, Bengaluru

Climate induced changes have had far reaching implications on agricultural production environment making life harder for vast number of poorest and most vulnerable population depending on it. Impact of adverse climate is felt most where poverty is widespread and agricultural and allied activities are marginalised. Concerted effort to overcoming restrictions imposed on available choices due to climate change through strengthening institutional innovations, technology generation and making available accurate information is need of hour. Farmers and stakeholders supported by vibrant and evolved institutions must respond to challenges posed by ubiquitous climate change. Historically, farmers and institutions have been remarkably successful adapting to a myriad challenges, and this should be facilitated in future as well.

There is a perceptible and sweeping change in climatic parameters world over, which inter alia, is affecting agriculture sector adversely. Frequency of occurrence of extreme weather events have increased significantly in recent years. Unfortunately certain components of agricultural sector acts as both cause and effect of climate change. In other words, climate change has a significant bearing on agricultural sector locally and in turn, some of agricultural practices effects climate change globally. In a sense, these two are mutually interdependent which has to be addressed holistically as it affects local communities. Intergovernmental Panel on Climate Change (IPCC) has reported that agriculture is responsible for over a quarter of total global greenhouse gas emissions. Given that agriculture's share in gross domestic product (GDP) of country of about 12 percent and 4 percent for world, suggests that agriculture is Greenhouse gas intensive. Thus agriculture can also be seen as part of problem.

This paper examines potential role of innovative agricultural practices and technologies in climate change mitigation and adaptation through institutional changes needed to encourage innovation and diffusion of these practices. Institutions are needed to moderate responses to climate change. These institutions should be supple to adapt swiftly to new demands imposed by weather variability from time to time. It is important to understand how on effective responses mechanism to climate change through innovation, viz., technological as well as institutional and relational can be institutionalised.

It is widely recognised that problems associated with climate change have to be addressed using place-based-approach rather than a universal approach. Mirroring this view, Yarnal (2010) and Moser (2001) outlined need for locality based research on vulnerability and adaptation.

Since there is livelihood dimension to climate change it is implied that climatic change research should address broader issues of intuitional arrangements and livelihood as well. The effect of climatic changes on agricultural sector is more pronounces on poorer sections like small and marginal farmers and landless labour because of climatic changes have potential to bring about long term changes in their resource endowment.

There are some really bleak projections of impact of climate change. The Stern report (Stern, 2007) has estimated that with a 2°C increase in average temperatures, world GDP would reduce by roughly 1% and World Development Report of World Bank (2009) has estimates that without adequate innovations to offset adverse effects of climate change annual GDP of India will reduce by a staggering 5%. Further, Cline in 2007 has estimated that regional disparities around global average impact are substantial. India and Africa are projected to see a reductions of agricultural output by 30% or more, which is a serious cause of concern. It is quite possible that many of these forecasts are pessimistic because they do not fully factor in possibility of adaptation, but nevertheless serves as a wakeup call for policy makers and stakeholders to act and not adopt a wait and watch policy.

To study coping strategies of climate change many approaches have been suggested. Ribot (1995, 2010), has suggested a very practical approach, "multiscale causal structure", of documenting local climate related problems and their impact. This approach helps in identifying both cause and how locally stakeholders have coped with it. The results so obtained could be examined spatially and insights into a range of strategies can be drawn up, which will facilitate a local and global understanding of problem. Thus, solutions to facilitate adaptation of agricultural sector to climate change have to be proactive and long term in approach, drawing heavily on three "Ins" of development of Paul Streeten, viz., Innovation, Institutions and Information.

Constraints in Adoption of Technology

Generation of climate relevant agricultural technologies and harnessing them to help farmers to adapt to changing climate will require innovations in institutions to deliver these technologies and receive feedback for continuous and proactive improvement. However impediments to development, diffusion and use of relevant technologies can surface at several levels–from inception of innovation to transfer of technologies and access to agricultural innovations by vulnerable smallholders.

Usually constraints manifest themselves at stage of adoption itself. The causes are many and these include poorly functioning markets for both inputs or outputs, weak institutions and institutional inertia, inadequate climate mitigating infrastructure, an extension system which is not oriented towards addressing problems related to climatic stress and a debilitated credit and insurance markets especially for poor.

A direct effect of climate change is scarcity or water. Therefore future of agriculture in these vulnerable regions hinges on improving efficiency of existing irrigation systems and, extending irrigation facilities and bringing dry areas under watershed development to improve in situ water conservation. Drip irrigation systems will play an important role on farmers' fields, and medium and large irrigation systems like canal irrigation systems will have to work more efficiently. In places with limited access to irrigation, watershed development should be practiced. Major innovations in response to climate variability will take form of improved information through local and global monitoring and forecasting.

A major problem that is facing generation of innovations with regard to public funded research, is paucity of funds to carry out relevant climate change related research. This general declining growth rates in public investments in agricultural R&D will constrains innovation of technologies related to climate change, as it is primarily in public good domain. Soon support of public agricultural R&D funds explicitly toward climate change mitigation and adaptation will need focus.

Absence of reliable and actionable information system about climate change and coping stratagems has become as complex predicament to farmers and other stakeholders and this aspect should be addressed and delivered on a real time basis. Historical patterns of climate are changing and hence relying on it for decision making is of limited use.

Institutions

Transition to a highly variable climatic does not occur in an institutional vacuum according to Agarwal (2008, 2010). For adaptation response to be effective, institution mechanism should be invigorated in such a way that it is responsive, flexible and scalable. Taylor *et al.* (2011) used hypothesis of induced innovation of Hayami and Rutten (1985), where scientists can view role of institutional innovations in driving coping strategies of a community. This hypothesis helps to explain how societies develop technologies that help them to cope with climatic stress, arbitrated by institutions. Thus vibrant institutions are needed. This would require partnership between stakeholders, principally Government and its parastatls, NGO's, Farmers, input suppliers, etc., at various levels which should essentially foster Public Private Partnership approach. However, an important insight into role of institutions, technological innovations and information is needed to help in coping with adverse effects of climate. Chhetri *et al.* (2010) avers that it is difficult test role of climate in stimulating institutions and technological innovations for improving adapting mechanism.

There is a need to institutionalise social safety nets which should include in its realm a vibrant insurance mechanism to help farmers in mitigating adverse effects of climatic stress. The experience with crop insurance in India is not encouraging from standpoint of sustainability, and perhaps it is true of all developing countries. It is supported by large government subsidies to farmers and insurance companies have been required to maintain widespread use of multi-peril insurance.

The challenge is to develop a sustainable insurance system that is self-supporting and raises incomes. Safety nets like direct finance transfers, interest rate subsidies, debt forgiveness have not been very effective globally due to inefficient targeting.

Innovations

The potential impacts of climate change on agricultural production in different parts of world have been assessed in numerous studies and reviewed. These studies range from major crops of region under study and emission scenarios have been simulated. Nevertheless, most studies indicate that agriculture in tropics is likely to be severely affected in coming decades by climate change. Some of key impacts on farming and food systems are documented by Vermeulen *et al.*, (2010). They have identified effects of climate change on crop yields, livestock, fish, biodiversity, pests and diseases both on crop and livestock, carbon fertilization, irrigation, food and fodder storage and distribution, food accessibility and utilization. To overcome adverse effects of climate change, off-setting innovations in agricultural sector are important and will be even more vital in context of climate change. Focussed innovations in climate sensitive crops have to be taken up on a continuous basis.

Use of induced innovation hypothesis put forth by Chettri *et al.*, 2012, underlines role of institutional innovation in helping stakeholders cope with climate stress through application of technology. Climate variability alters resource base in a location, irreversibly at times. This requires new technology production to adapt to changed scenario as well as an improved knowledge system to provide valuable information support. The development of technology in this framework should be facilitated by innovative institutions. This has been adapted and depicted in Fig. 1. The figure adapted from Chettri *et al.* (2012) shows role of institutions, innovations and information in ameliorating harmful effects of climate change.

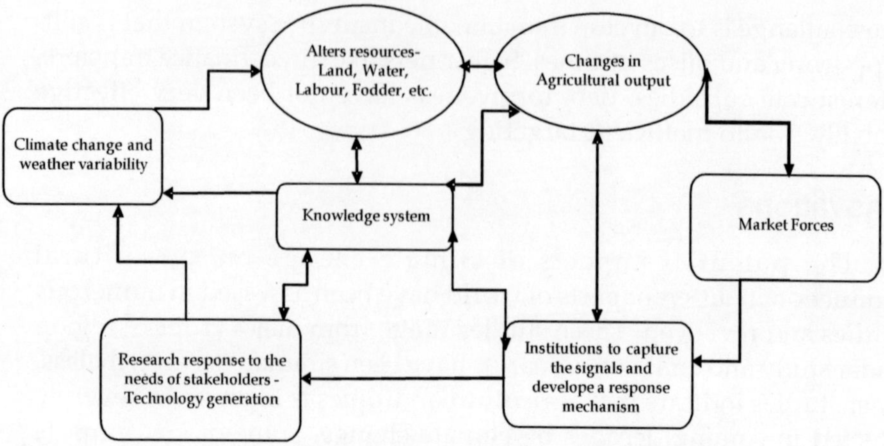

Fig. 1. Institutional, Technological and Information response to Climate change

Improving knowledge system: databases and models

As farmers deal with maze of climatic parameters and variability in weather, historical evidence becomes less reliable. There is a need to improve forecast weather events and weather probabilities and projection of likely effects on crop and livestock production, to aid their decision making process. Due to erratic nature of critical components of weather, timely and accurate information will have a very important role to keep stake holders informed about developments.

Improved micro-climate modelling and its dissemination can go a long way in helping farmers cope with problem of adverse effects. This could lead to improved short term forecasts, which could be disseminated via SMS. Lastly, better and more timely information can also help to forecast impending weather events such as scarcity of rainfall more effectively and thereby improve response times and adaptation (Mude *et al.*, 2009).

Forecasting climate change is complex, due to widespread variations and intricate factors affecting it, most of which are not clearly understood. This is leading to changes in precipitation coupled with increasingly erratic intra-annual and complex spatial different weather patterns. Besides weather forecasting, per se, its implications should be interwoven with likely changes it could have on agriculture fabric of region and should be developed as an "app' for ready and quick access and supply necessary information needs of farmers.

Whether a particular zone expects to become wetter or drier on average in coming decades, water management is central to farmers' adaptation to climate change.

Technology has greatly improve quality and accessibility of data on agriculture ant this could be modified to collect related to climate change. Accessing and using remote sensing of weather information and relating them to validation of different land use changes and dissemination of information using mobile phone technology is possible. But many of this information need to be validated on field based on land-based observations before they are put to use.

Summary

Climate change has become a reality and it is affecting people of all walks of life and directly on agricultural production and livelihoods. Climate is one of most important factors that farmers in country have to adjust to. The frequency of climatic variability and extreme weather conditions has made need for a response very compelling. While causes of climate change are not clearly understood to take corrective action for mitigating its adverse effects is now being pursued vigorously. A systematic effort to study climate change should be attempted following multi-stage causal structure at a micro level to understand possible causes and coping strategies that have succeeded in order to build a strong response mechanism. Unfortunately most affected are poor and marginalised areas, which calls for urgent steps from Government. Lack of data and its monitoring is a major limitation in establishing an empirical relationship between climate and its causes. This needs to be strengthened. Thus, a strong and effective response mechanism to adversity of climate change will have to bring together institutions, technological innovations and an improved knowledge system which will deal with data and models to inform so that coping mechanism can be more effective.

Reference

Agrawal, A. (2008). *The role of local institutions in adaptation to climate change.* In Paper prepared for social dimensions of climate change. The World Bank.

Agrawal, A. (2010). Local institutions and adaptation to climate change. In R. Mearns, & A. Norton (Eds.), *Social dimensions of climate change: Equity and vulnerability inthe warming world.* TheWorld Bank. Pp. 173-198.

Chhetri, N. and Sundar S. S. (2004). *The Prospects of Agricultural Adaptation to Climate Change: Climate-Technology Interaction in Rice -Wheat Cropping System in Nepal.* Paper prepared for presentation at American Agricultural Economics Association Annual Meeting, Denver, Colorado.

Chhetri, N., Easterling, W. E., Terando, A. and Mearns, L. (2010). Modeling path dependence in agricultural adaptation to climate variability and change. *Annals of Association of American Geographers.* 100(4): 894-907.

Chhetria, N., Chaudhary, B. P., Tiwari, P. R. and Yadawd, R. B. (2012). Institutional and technological innovation: Understanding agricultural adaptation to climate change in Nepal. *Applied Geography.* 33:142–150

Challinor, A.J., Wheeler, T., Garforth, C., Craufurd, P. and Kassam, A. (2007). Assessing vulnerabilityof food crop systems in Africa to climate change. *Climatic Change.* 83: 381-399.

Cline, W.R. (2007). *Global Warming and Agriculture: Impact Estimates by Country.* Peterson Institute for International Economics.

Hayami, Y. and Ruttan, V. W. (1985). *Agricultural development: An international perspective.* Baltimore: The John Hopkins University Press.

Lybbert, T. and Sumner, D. (2010). *Agricultural Technologies for Climate Change Mitigation and Adaptation in Developing Countries: Policy Options for Innovation and Technology Diffusion.* International Food and Agricultural Trade Policy Council, International Centre for Trade and Sustainable Development Issue Brief No. 6.

Moser, S. C. (2010). Now more than ever: need for more societally relevant research on vulnerability and adaptation to climate change. *Applied Geography.* 30(4): 464-474.

Mude, A.G., Barrett, C.B., McPeak, J.G., Kaitho, R., and Kristjanson, P. (2009). Empirical Forecasting of Slow-Onset Disasters for Improved Emergency Response: An Application to Kenya's Arid North. *Food Policy.* 34(4):329-339.

Ribot, J. C. (1995). The causal structure of vulnerability: its application to climateimpact analysis. *GeoJournal.* 35(2): 119-122.

Ribot, J. C. (2010). Vulnerability does not fall from sky: toward multi-scale propoor climate policy. In R. Mearns, & A. Norton (Eds.). *Social dimensions of climatechange: Equity and vulnerability in a warming world.* Pp. 47-74.

Taylor, R.D., Mette. F.O. and Chhetri. N. (2011). Adaptation as innovation, innovation as adaptation: An institutional approach to climate change. *Applied Geography*. xxx. Pp.1-5.

Stern, N.H. (2007). The Economics of Climate Change: The Stern Review. Cambridge University Press,World Bank.

Vermeulen, S.J., Aggarwal, P.K., Ainslie, A., Angelone, C., Campbell, B. M., Challinor, A. J., Hansen, J.W., Ingram, J.S.I., Jarvis, A., Kristjanson, P., Lau, C., Nelson, G.C., Thornton, P. K. and Wollenberg, E. (2010). *Agriculture, Food Security and Climate Change:Outlook for Knowledge, Tools and Action.* Background paper prepared for The Hague Conference on Agriculture, Food Security and Climate Change on behalf of CGIAR by Program on Climate Change, Agriculture and Food Security ofthe Consultative Group on International Agricultural Research (CGIAR) and Earth System SciencePartnership (ESSP).

Yarnal, B. (2010). Climate change and applied geography place, policy, and practice: a foreword to special issue. *Applied Geography*. 30(4): 461-463.

World Bank. (2009). *Development and Climate Change*. World Bank.

□□□

Chapter-17

Information and communications technology in improving small farm productivity

M.V.ASHOK

Chief General Manager, NABARD, Mumbai

The first green revolution of nineteenth century is over. The next green revolution in India is generation of technology and infrastructural development. Indian agriculture is confronted with some major issues and challenges that continue to hinder growth of this sector. The challenge for government and policymakers is to 'regain agricultural dynamism'. The country needs a strong pull-up support to agricultural sector, which should grow at rate of at least 4 percent per annum. The major dilemma that agriculture sector faces under situation of recent global food crisis and rising food prices is of striking a balance between policies of food security and improving income levels of farmers.

Along with this, major challenge India is facing is that India's average operational landholding is less than two hectares, with a majority of farmers being small or marginal landholders. Improving livelihoods of small farmers has been a cornerstone of Indian government policy targets for many years and is imperative for social and economic development.

At farmers' level, sustainability concerns are being expressed that input levels have to be continuously increased to maintain yields at old level. This poses a threat to economic viability and sustainability of crop production. Research, extension, literacy and infrastructure have been identified as most important sources of growth in farm productivity. Development of markets improves input/output market interface and it is of crucial importance for growth in farm productivity. Human resource development is central to adoption of technology and promotion of sustainable development.

Small and Fragmented Land-holdings

The seemingly abundance of net sown area and total cropped area of) pales into insignificance when we see that it is divided into economically unviable small and scattered holdings. The average size of holdings was 2.28 hectares in 1970-71, which was reduced to 1.82 hectares in 1980-81 and 1.50 hectares in 1995-96. The size of holdings will further decrease with infinite Sub-division of land holdings. The problem of small and fragmented holdings is more serious in densely populated and intensively cultivated states like Kerala, West Bengal, Bihar and eastern part of Uttar Pradesh where average size of land holdings is less than one hectare and in certain parts it is less than even 0.5 hectare. The main reason for this sad state of affairs is our inheritance laws. The land belonging to father is equally distributed among his sons. Sub-division and fragmentation of holdings is one of main causes of our low agricultural productivity and backward state of our agriculture. A lot of time and labor is wasted in moving seeds, manure, implements and cattle from one piece of land to another. Irrigation becomes difficult and costly on such small and fragmented fields. Further, a lot of fertile agricultural land is wasted in providing boundaries. Under such circumstances, farmer cannot concentrate on improvement.

Thus, feature of small and marginal farmers dominating multitude of farming families is something, which cannot be done away with. When these small farmers operate on a stand above basis they are faced with a plethora of problems.

i) Lack of information for farmers

Lack of extension service for small farmers in terms of weather conditions, moisture content of soil, taking up appropriate farming activities as per condition of crop and weather conditions is a major challenge for farmers. Although extension service is responsibility of state governments, quality of service is so poor in most of states that farmers tend to depend on either themselves or other farmers.

ii) Irrigation

Although India issecond largest irrigated country of world after China, only one-third of cropped area is under irrigation. Irrigation is most important agricultural input in a tropical monsoon country like

India where rainfall is uncertain, unreliable and erratic India cannot achieve sustained progress in agriculture unless and until more than half of cropped area is brought under assured irrigation. It is small farmers who are largely affected due to this problem and it is this class of farmers who are mostly depended on monsoon.

iii) Lack of mechanization

In spite of large scale mechanization of agriculture in some parts of country where land holding size is large, most of agricultural operations in larger parts where land holding size is small are carried on by human hand using simple and conventional tools and implements like wooden plough, sickle, etc. Little or no use of machines is made in ploughing, sowing, irrigating, thinning and pruning, weeding, harvesting threshing and transporting crops. This is specially case with small and marginal farmers. It results in huge wastage of human labor and in low yields per capita labor force. There is urgent need to mechanize agricultural operations so that wastage of labor force is avoided and farming is made convenient and efficient. Agricultural implements and machinery are a crucial input for efficient and timely agricultural operations, facilitating multiple cropping and thereby increasing production.

iv) Agricultural marketing

Agricultural marketing still continues to be in a bad shape in rural India. In absence of sound marketing facilities, farmers have to depend upon local traders and middlemen for disposal of their farm produce, which is sold at throw away price. In most cases, small farmers are forced, under socio-economic conditions, to carry on distress sale of their produce. In most of small villages, farmers sell their produce to moneylender from whom they usually borrow money. Further, saleable produce with small farmers being a small quantity mostly does not form marketable lot and is therefore forced to sell to traders who aggregate to form marketable lots which fetches better price. In absence of an organized marketing structure, private traders and middlemen dominate marketing and trading of agricultural produce. In order to save farmer from clutches of money-lenders and middle men, government has come out with regulated markets. These markets generally introduce a system of competitive buying, help in eradicating malpractices, ensure use of standardized weights and measures and evolve suitable machinery for settlement of disputes thereby ensuring that pro-ducers are not subjected to exploitation and receive remunerative prices.

v) Inadequate storage facilities

Storage facilities in rural areas are either totally absent or grossly inadequate. Under such conditions farmers are compelled to sell their produce immediately after harvest at prevailing market prices, which are bound to be low. Such distress sale deprives farmers of their legitimate income.At present there are number of agencies engaged in warehousing and storage activities. The Food Corporation of India (FCI), Central Warehousing Corporation (CWC) and State Warehousing Corporation are among principal agencies engaged in this task, which are available for non-perishable produce like grains. For farmers cultivating perishable produce like vegetables and fruits absence of cold storages is a major challenge.

vi) Inadequate transport

One of main handicaps with Indian agriculture is lack of cheap and efficient means of transportation. Even at present there are lakhs of villages, which are not well connected with main roads or with market centres. Most roads in rural areas are Kutcha (bullock- cart roads) and become useless in rainy season. Under these circumstances, farmers cannot carry their produce to main market and are forced to sell it in local market at low price. Linking each village by metaled road is a gigantic task and it needs huge sums of money to complete this task. Small farmers having small quantity having small quantity of produce fund it extremely expensive to transport their produce individually.

Advent of Information Communication Technology in Farming

In agriculture, education creates conditions that enable farmers to acquire and use knowledge for decision-making regarding allocative and technical matters effectively. Information-based, decision-making agricultural system, otherwise known as Precision Agriculture, is designed to maximize agricultural production and is often described as next great evolution in agriculture. The combination of Global Positioning System (GPS) and mobile mapping are supposed to provide farmers with information for implementation of decision-based Precision Agriculture. In context of India, use of mobile phones as a mode of providing agriculture related information would depend on how far mobile phone network has been able to link farmers to market information - timely and accurately. The impact on productivity can be directly measured in terms of increased returns to farmers with a trickle-down effect on cropping pattern and potential yield of sown crop. Information on price factors-prices of inputs and output, and non-price factors like information about availability of inputs, quality of seeds, modern techniques, etc. would play primary role in improving farm productivity. The increasing penetration of mobile phone networks and handsets in India therefore presents an opportunity to make useful information available more widely. The recent introduction of a number of mobile-enabled information services suggests that it is timely to take a look at their impact on agriculture sector in India. These services deliver a wide range of information to farmers.

The impact of mobile phones is reflected across Indian agriculture, particularly for small farmers. Mobile phones would help in reducing informational asymmetry existing in agricultural sector and would be helpful in improving farm productivity and profitability. There would be a positive impact on farmers' profitability by reduction in

i) Transaction costs at both input and output levels;

ii) Search costs through time saving; and

iii) Travel cost.

A better and timely decision-making on cropping pattern could also increase farm profits. The use of superior quality inputs, mainly seeds would also deliver better yields and profits. The crucial sense behind the study was that information received through mobile phones could play a complementary role to extension activities and would have a better impact than other one-way information technologies.

Information and Communication Technology (ICT) consists of three main technologies. They are:

1. Computer Technology
2. Communication Technology and
3. Information Management Technology.

These technologies are applied for processing, exchanging and managing data, information and knowledge. Despite significance of agriculture sector in India, due to various challenges posing sector it is declining. Initiatives to enhance agricultural productivity, competitiveness and rural growth are being taken. Introduction of Information and Communications technology (ICT) has emerged as a major source for development in India, and especially in agriculture sector. ICT enables smooth dissemination of required information at the right time, which is crucial to improve productivity, plan better, and therefore deliver better results. This revolution in information technology has made access to information easy and cost-effective.

Though ICT has a huge potential, ICT models have certain barriers. There is a huge digital divide between urban and rural population. Most of the ICT models are based on internet and computers, which are not accessible to rural population due to lowincome levels and high levels of illiteracy. Along with limits regular reach to majority because various reasons like insufficient power availability in rural areas, poor ICT infrastructure, ICT illiteracy, non-availability of timely relevant content, non-integration of services, poor advisory services and lack of localization, and in particular non availability of agricultural information kiosks/ knowledge centers at grass root level. To bridge this gap options of voice based communications technology of mobile and community radios are powerful, feasible and effective mediums. Especially telecommunications is transforming rural agriculture space rapidly. It plays a vital role in contributing to empowerment of farmers living in villages. Use of Mobile technologies as a tool of intervention in agriculture is increasingly popular. Smartphone penetration enhance multi-dimensional positive impact on sustainable poverty reduction and identify accessibility as main challenge in harnessing full potential in agricultural space. Reach of smart phone even in rural areas extended ICT services beyond simple voice or text messages. Several smartphone apps are available for agriculture, horticulture, animal husbandry and farm machinery.

Role of ICT in Agricultural Production

There are mainly three roles played by ICTs in agricultural production. They are:

1. **Enhancing Agricultural Production**

   Farmers (especially smallholders) often face threats from many threats like poor soils, drought, erosion and pests. Key areas where ICT can help improve this is by providing up-to-date information about pest and disease control, early warning systems, new varieties, new ways to optimize production and regulations for quality control.

2. **Improving Market Access**

   Providing up-to-date information on market prices of commodities, inputs and consumer trends. This can improve a farmer's negotiating position and their livelihood, while enabling farmers to make better decisions about future crops and commodities, and also best time and place to sell and buy goods.

3. **Capacity-building and empowerment**

   ICT technologies can be used to strengthen communities and farmer organizations strengthen their own capacities and better represent their constituencies when negotiating input and output prices, land claims, resource rights and infrastructure projects. Rural communities are able to interact with others via use of ICT, which reduces social isolation that they would otherwise be facing. Besides that, ICT technologies are able to make processes like law-making and land title approvals more transparent

Use of Mobile Phones a Boon for Agriculture

Mobile phones have potential to provide solution to existing information asymmetry in various lagging sectors like agriculture. India's agricultural sector suffers from low growth rates and low productivity. Issues in access to information are weak points at every stage of agri-supply chain. For small farmer-based economy like India, access to information can possibly enable better incomes and productivity to farmers. While Mobile phones can act as catalyst to improving farm productivity and rural incomes, quality of information, timeliness of information and trustworthiness of information are three important aspects that have to be delivered to farmers to meet their needs and expectations. There exist critical binding constraints that restrict ability of farming community to realize full-potential gains and it is more so for small than large farmers.

Mobile phones are omnipresent and cost effective means to revolutionize agriculture in India. According to Government of India census 2011, mobile penetration in rural India stands at an astounding 51 percent. Mobile phones therefore can be effectively utilized for myriad purposes including generating, processing, transmitting, disseminating, sorting, archiving and retrieving critical information and data.

India has seen several ICT related interventions helping farmers and particularly small farmers in their operations. A few such interventions are captured below to give an idea of range of services and benefit it accrues to small farmers.

i) ITC e-choupal

ITC e-choupal effort is driven by a company involved in grain trading 4000 Internet choupals (kiosks) in villages to aid grain procurement, support agriculture and grain trading.

ii) Jayalaxmi Agrotech service

Jayalaxmi Agrotech Pvt Ltd, a rural social entrepreneurship start-up firm from Karnataka bridges information gap using ICT solutions with its dedicated apps reaching one farmer every minute. These "Make in India" apps from social entrepreneurs from Karnataka have already won "Mobile for social good" award and "mBillionth award" for mobile solutions. Jayalaxmi agrotech has been a finalist for "Digital India Challenge". Jayalaxmi agrotech is in spirit of "Digital India" and "Make in India". Government of Karnataka and NABARD Bangalore RO have partnered with agriculture ICT solutions provider to enhance its coverage among farmers.

iii) Village Knowledge Centers (VKCs)

Village knowledge centers of MS Swaminathan research foundation were launched in 1998 in Pondicherry. The main aim behind establishment of VKCs was to provide sustainable food security in rural areas of Pondicherry. To fulfill this aim, it provides technical information related to agricultural inputs. It helps in procuring quality seeds, in providing information about daily marked priced from government as well as private bodies, and advices farmers on rotation of crops as well as about use of fertilizers and pesticides. VKCs receive information by voice mail, and disseminate it through any public address system. It has also identified 13 districts in Pondicherry, where there is a huge potential for agriculture business, and where government will invest ₹ 170 cr.

iv) Digital Green

Digital Greenis an NGO working to improve social, economic, and environmental sustainability of small farmers in India has developed innovative ways to improve Indian agriculture system. They have pioneered a novel process of creating local videos of farmers speaking on agriculture practices in their area. Their theory of change is that diffusion of best practices happens faster when videos are more contextual, or viewers can relate to them, for example by seeing their neighboring farmer as being featured in video. In partnership with NGOs, Digital Green screens these videos using pico-projectors in small groups of 6-10 farmers. Together with Ford Foundation, they did a small project on using mobile phones with projectors that were given to a few key people in each village, and played videos to friends and associates of people.

v) CROPIN transforms Indian agriculture by using cloud

Using a cloud-based platform, startup Crop In Technology is aiming to empower Indian farmer by making every crop traceable so that harvested crops meet best global agricultural practices. While Indian agriculture has been single largest source of employment in India, there is a big mismatch, as yield does not match global standards. India has a yield rate of 2.4 tons per hectare, compared to China, which has nearly double yield at 4.7 tons per hectare. Startup CropIn Technology wants to change this scenario through its vision of digitizing agriculture by using a combination of cloud-based technologies and mobile phones to improve farm yield. CropIn Technology was established in August 2010 to develop and offer on a low cost pay model as you use product on an IT platform on cloud integrated with a Windows or Android-based smart mobile app. The vision was to bring affordable state of art technology in agriculture and mission was to make every farm visible online, make every farmer adopt best global agricultural practices and to make every crop traceable so that harvested crops meet global quality standards and thus become export worthy. This technology makes an impact on livelihood of farmers at bottom of pyramid by generating more productivity out of same farm. By using a mobile app developed by CropIn Technology, a surveyor can input details related to a farm and monitor efficiency of a farm. Crops are regularly monitored on a thorough basis with respect to fertilizer sprayed and detection of pests, if any. Regular monitoring of fertilizers is essential to ensure that crops exported are compliant with required safety levels of different chemicals as prescribed by different countries for consumption. The mobile app captures information related to farm surveyed and relays it back to a cloud-based server.

vi) e- Arik (e-Agriculture for educating farmers in North East)

The Village Knowledge Centers in North East India is one such initiative which uses ICT as an educational tool. An ICT based initiative entitled e- Arik (e-Agriculture) was initiated during mid 2007. The first Village Knowledge Centre (VKC) was set up in Yagrung village in East Siang district of Arunachal Pradesh State. There is project team comprising of Research fellows and computer instructor, who serve as facilitators helping farmers use VKC. Farmers need to register themselves to be able to use services. Presently, 500 farmers from 12 remote tribal villages are registered under e-Arik system. The project team helps farmers access ICT based information. They also help them in engaging in remote consultation with other agricultural experts. Both Central Agricultural University (CAU) and Krishi Vigyan Kendra provide technical support. While CAU is responsible for overall project implementation and to provide scientific advisory services to farmers, KVK has been organising trainings and demonstrations on farm and carried out follow up activities. The Project Advisory Committee comprising of project team, village tribal council members and farmers, regularly monitors, supervises and is responsible for quality control of project activities.

vii) RainbowAgri

Rainbow Agri, has a host of services where farmers can talk to communities, users and farmers can buy and sell local products and buyers can even get live price updates. One can even connect with neighborhood farmers directly and see real images of their crop. Rainbow Sell gets farmers, street vendors and even small stores to reach buyers in their local areas and even customize deliveries to suit buyer's needs, by holding provisions for discounts, enter details in local language and even track status of order. With Rainbow Live

Price, all crops in India are arranged by location, weather, and market names and of course type and pretty comprehensive. It is often called internet of farmers.

viii) MandiTrades

Mandi Trades enables farmers to get information about commodity prices directly provided by Government of India, produce info and a map based view of available farm products with ability of connecting to a required farmer via just a simple phone call through this app. A buy/sell platform aims to solve problems of marketing and farmers can avoid daily struggle to sell their produce. When they register with site, it captures all of their produce and location details and updates its database. Farmers can know of average demand, peak/low demand season, in and off-season needs, rare items and their prices and can plan their farms accordingly.

ix) CropLense

CropLense is a field enablement tool that helps farmers address crop issues in a cost-effective and timely fashion. The app helps person on field (farmer or field officer) capture high-resolution images of crops that need protection, annotate images with voice messages, and then avail expert advice from farm consultants, entomologists or soil experts. CropLense includes an online forum where enterprises, government agencies, educational institutions, students and farmers can post queries, seek clarifications and interact with peers and experts.

Key Features

- **Gain information with any handheld device:** Farmers and field officers can gain information on crop-based issues and on-field problems on any handheld device with just an internet connection.
- **High-resolution pictures and voice annotations:** Farmers can attach high-resolution pictures (taken with a 3MP or higher camera), and add voice annotations to avoid misinterpretation of data, thereby getting accurate timely advice from experts.
- **Access to multiple responses:** The app allows farmers to gain access to multiple responses or points of view in real time, from experts around the world.

- **Expert agricultural advice worldwide:** Farmers can gain access to agricultural experts and advisors worldwide.
- **Voice messages in local languages:** The app can record voice messages in preferred local language.
- **Instant alerts:** After a query has been raised on any crop-related issue, tool provides instant notification via SMS alerts when issue is resolved.
- **Multiple images upload:** The app can upload multiple images at same time for a single query.
- **Platform-agnostic tool:** Farmers can now easily report and view expert advice on mobiles, computers, tablets, or any handheld device.

Benefits

- The technology gets accurate and effective advice on resolution of crop related issues.
- Farmers get access to multiple experts and advisors, thereby resolving crop issues on time.
- Farmers as well as Field Officers save costs and travel time.
- Very importantly, analytical insights generated through app usage patterns from farmers will be game changer.
- All these apps are offline and can work without internet. Basic android smart-phone is sufficient to use these applications. Beneficiary farmers can transfer apps to other farmer's phone through Bluetooth or share it.

x) MyYara

Water scarcity has become one of biggest challenges we need to face. In Asia climate change is threatening water supply of one billion people. It is, therefore, important to embrace new irrigation technologies to reduce water waste. Agriculture consumes about 70 percent of all freshwater withdrawals, and will have to do with less in future. The MyYara technology contains a sensor with a magnet, which is placed on a leaf on plant and measures water pressures in leaf between two magnets. Based on scientific research, it will be possible to interpret water pressure to indicate if plant needs more or less water. The sensor is connected to a transmitter, which transfers data to our cloud-based solution (MyYara). It is also connected to weather forecasts; so that plants are not watered if it will rain in near future. This ensures that water is only applied when needed, which may lead to major savings.

xi) Reuters Market light (RML)

Reuters Market Light (RML) is a unique ICT model compatible to all mobile service providers, which delivers mobile-based information service for farmers. It was launched in October 2007 and it provides individual farmers with customized, localized and personalized weather forecasts, local crop prices, agricultural news and relevant information (i.e., crop advisory) in the form of SMS messages sent to their mobile phones in their local language. This allows subscribing farmers to plan their sowing, irrigation, application of fertilizers, and harvest – thus helping them manage some of their risks, as well as to decide when and where to sell their produce to maximize profit.

RML has built an end-to-end value chain from sourcing through to rural distribution to meet decision-critical needs of farming sector over mobile phones. It currently covers over 50 commodities and over 250 markets in two states. RML provides following information set to its subscribing farmers:

- Spot Crop Prices on a daily basis (2 crops and 3 markets)
- Localized *taluka*-level (50 km radius) weather forecast
- Crop advisory
- Commodity news & information
- Any other agri-rural information relevant for farming community

xii) Mpower Social

Livestock farming is one of most important forms of livelihood for many families. Unfortunately ratio of registered veterinarians to cattle is very low and Mpower's ICT Livestock Management Systems (LMS) aims to decrease this gap by connecting livestock professionals to smallholder farmers via Livestock intermediaries. These Livestock workers register farmers, gather regular updates about their cattle and provide expert medical advice-all through their mobile application. This reduces cost of consulting an expert or a doctor.

xiii) Fisher Friend Mobile Application

Fisher Friend Mobile Application (FFMA), which gives fishermen a wealth of information from exact geographic co-ordinates of shoals of fish, real-time data on wave length and wind speed. Around 10,000 fishermen families across Tamil Nadu, Andhra Pradesh and Kerala are using the app, which was piloted by MS Swaminathan Research Foundation (MSSRF) in partnership with Qualcomm and TCS in 2013. In addition to information on movement of fishing shoals, which is updated once every 24 hours, app also marks danger zones like coral reefs and sunken ship wreckages. Other details like wind speed and wave height are updated once every 6 hours. Fishermen say that app has boosted their catch and thereby incomes. "Previously I used to fish 6 to 10 tons of tuna, now with help of information, I am able to catch up to 20 tons," said Chennai-based fisherman K Alex. The app has also reduced travel time and expenses, say fishermen. The technology for tracking fish shoal movement is simple. Fish shoals tend to congregate around waters that have vast colonies of phyto-planktons, which are basically microscopic algae. The Indian National Centre for Ocean Information Services (INCOIS)—an autonomous body under ministry of earth sciences which tracks everything from potential tsunamis to wave forecasts-captures satellite images and passes them onto MSSRF which in turn processes and uploads information to app. The fishermen usually use data from app to set course before they set out on a fishing expedition. MSSRF officials say that tests conducted by them reveal that mobile reach usually extends up to 25 km into sea. A pilot to boost network reach to as much as 100 km into sea is on anvil in Kollam in Kerala. "We have also helped MSSRF in training fishermen to use Android app," said Anirban Mukerji, senior manager government affairs, Qualcomm, adding that company helped integrate Google analytics onto app to get information about usage data for all the features on the app. Another popular feature of app is an alarm that starts beeping as soon as a boat approaches international water boundaries. In the past, fishermen from Tamil Nadu have been arrested or worse, shot at, when they have strayed into Sri Lankan waters.

xiv) mKisan

mKisan was launched in 2013 and gives information, services, and advisories to farmers by SMS; farmers can indicate their preferred language and specify relevant agricultural practices and data specific to their location. mKisan has sent millions of messages throughout length and breadth of India.

xv) Kisan Call Centers

Kisan Call Centers were launched in 2004 over the entire country to deliver extension services to farming community. Farmers can call and make queries related to crops, seeds, fertilizers, agricultural commodity prices, pesticides, horticulture, veterinary issues and so on, free of charge.

xvi) Green Sim

Green Sim another private national initiative, is a joint venture by IFFCO (Indian Farmers FertilizerCooperative) and Airtel, fulfilling their individual objectives of both telecom provider and cooperative by providing information directly to farmers, delivering location-and time-sensitive information, and important alerts without adding economic strain to farmers. Green Sim comes with a flat cost of 86 rupees and can get IFFCO services free of charge with proactive messaging services which target at updating of new information in market, weather and cropping patterns, and reactive messaging services which are need-based.

xvii) iKisan

iKisan was established in 2000 as part of Nagarjuna Group's vision of rural prosperity through enhanced knowledge, with technology making Indian farmers globally competitive. iKisan is an agricultural portal which provides online, detailed content on crops, crop management techniques, fertilizers, and pesticides and a lot of other agriculture-related material. Other initiatives by private sectors entities discussed included mKrishi, a mobile agro-advisory system by Tata Consultancy Services (TCS).

xviii) e-Sagu

Other forms of online portals are also available such as eSagu personalized agro-advisory system, use of Facebook by Kerala's Department of Agriculture, and SasyaSree. eSagu is an IT-based scalable system to disseminate location-specific best agricultural practices to all farmers; an agricultural scientist provides expert advice using digital photographs of crop.Facebook is being widely used by Department of Agriculture, Kerala in state's local language as a channel to disseminate useful information to farmers and facilitate communication among famers. SasyaSree, managed by Centre for Good Governance (CGG), is another local language portal (this one in Telugu) which promotes use of existing technologies as means to improve productivity and farmer incomes. It is currently available in eight districts of state, providing informative videos based on local conditions and local cropping patterns.

xix) Rice Knowledge Management Portal (RKMP)

The Rice Knowledge Management Portal (RKMP) is largest knowledge portal for rice in the world. Rice researchers across world can access, share, and discuss research data, themes, tools, and knowledge for better rice science. It also provides access to trade information, food safety regulations, mandi market prices in addition to regular queries related to varieties, diseases, and pests etc. It also supports diagnostic tools such as disease/pest diagnostic phone apps.

xx) Life Line Agriculture Program

One World, based in Delhi, helps farmers with what they call "Life Line Agriculture Program": farmers can call a toll-free number for online advisories such as disease attack, or sudden emergencies such as livestock disease based on specialist information that they can put to practical use. It has reached 150,000 farmers in 1000 villages across five districts in Uttar Pradesh. The system continuously compiles a list of frequently asked questions; there are currently 490,000 questions and answers, and this list grows every day.

xxi) EM 3 Agricultural Services

EM3 AgriServices Pvt. Ltd. provides farmer with a suite of technology-based services that will give farmer comprehensive solutions to improve his productivity through their brand "SAMADHAN", which is a network of farm service centers- "Samadhan Kendras". Each Kendra (means a centre,a focal point), managed through robust processes, IT enabled systems, and manned by agronomy and agriculture technology professionals, is equipped to deliver entire suite of basic and precision operations throughout farming value chain. This include, soil analysis, seedbed preparation, sowing, fertilizer application, weed/pest control, top dressing, harvesting and post-harvest operations.

The services will vary by crop and by geography, but fundamental principles remain constant. These principles are designed to benefit farmer and are:

- A one-stop shop for farmer's needs.
- High quality and state of art equipment–to create highest possible value for farmer.
- Customized services, literally tailored for every individual farm.
- Timely–every activity for farmer is critical. We configure ourselves appropriately.
- Reasonably priced; far below manual cost.

xxii) mKrishi

mKRISHI® developed by Tata Communications Sercives (TCS)provides farmers with benefits of Information and Communications Technology (ICT). It is an award-winning, patent-pending innovative platform that delivers services to rural communities. It connects farmers with a variety of stakeholders packaging multiple services through communication devices like mobile phones. It can also integrate wireless sensors and script technology with communication devices to provide an enhanced solution. **mKRISHI®** ensures significant business benefits to stakeholders by enabling them to reach farmers directly. TCS' **mKRISHI®** platform uses mobile technology to cater to absolute needs of rural sector. It offers personalized advisory services in voice and visual using communication devices like mobile phones.

mKRISHI® serves to achieve following:

- Reach farmers individually to understand their needs.
- Enable farmers to receive important information about pesticides, fertilizers and soil and water conservation.
- Provide critical microclimate/weather information in order to plan farming operations.
- Facilitate better production and cultivation practices.
- Improve access to markets, refining associated economy.

Some of Important Efforts by Public Sector, NGOs and Private Sector in ICT for Farming are

- Cyber extension program of MANAGE (http://www.manage.gov.in/)
- National Informatics Centre (NIC) (http://home.nic.in)
- DACNET (http://dacnet.nic.in)
- Community Information Centre (CIC) (http://www.cic.nic.in/)
- Warana Wired Village Project (http://www.mah.nic.in/warana)
- Bhoomi (http://www.bhoomi.kar.nic.in/
- e-Seva (http://www.esevaonline.com)
- http://ruralinformatics.nic.in/
- N-Logue- http://www.n-logue.co.in/home.htm
- http://edevelopment.media.mit.edu/SARI/mainsari.html
- http://www.tenet.res.in/cordect/cordect.html
- http://www.drishtee.com/
- http://www.simputer.org/
- http://www.lokmitra.gov.in
- http://www.gyandoot.net/
- http://himachal.nic.in/lokmitra/lokmitra_hdr.htm
- http://setu.maharashtra.gov.in/
- http://www.agmarketnet.nic.in
- http://www.itcportal.com/sets/agreeex_frameset.htm
- http://www.oddanchatrammaket.com/
- http://www.agriwatch.com
- http://indiagriline.com/
- http://www.ikisan.com
- http://www.indiagriline.com/
- http://www.tatatkk.com/kisan_kendra.htm

- http://www.nddb.org/
- http://www.isapindia.org/
- http://www.mssrf.org/
- http://www.sristi.org/
- http://www.icar.org.in
- http://gauanand.com/
- http://raitumitra.kar.nic.in/

Reference

Agrawal, B. (2006). Communication Technologies and Rural Development in India: Promises and Performances, Indian Media Studies. In the 2014 WEI International Academic Conference Proceedings Budapest, Hungary The West East Institute.

Anandaraja, N. (2008). e- Extension Initiatives of TNAU. *Information Technology for Growth in Agriculture.*

Case-study-1-IFAD-ICT-India.pdf Center for e-Governance, Indian Institute of Management. (2010). "An Evaluation of Gyandoot. Ahmedabad." http://www1.worldbank.org/publicsector/bnpp/Gyandoot.PDF

Bhatnagar,F.S. (2008), ITC e-Choupal: Corporate Social Responsibility in Rural Indiahttp://seasofchange.net/wp-content/uploads/downloads/2012/03/

Gupta, S.D. (2006). How IT is changing rural India. Rediff.com news, April07, 2006.

UNWTO Report. (2012). UN e-Government Report

Kuhlmann, F. (2005). IT Applications in Agriculture: Some Developments and Perspectives.

Saravanan, R. (2007). ICT Initiatives for Agriculture Extension in North-East India.

Senthilkumaran, S., & Arunachalam, S. (2002). Expanding Village Knowledge Centers in Pondicherry. Regional Development Dialogue

Zaleds, S.R. and Zaleds, S.M.(2012). Role of Information and Communication Technologies in modern Agriculture. *International Journal of Agriculture and Crop Sciences.*

Srinivas, J. (2004). Sustainable Access in Rural India (SARI) Project report. www.iitb.ac.in/ICTforD/SARI%20final.pdf

Ahuja,V.(2011). Cyber Extension: A Convergence of ICT and Agricultural Development. *Global media Journal*.

Zjip, W. (1994) Improving Transfer and Use of Agricultural Information: A Guide to Information Technology. *World Bank Discussion Paper 247*. Washington DC: World Bank.

□□□

Chapter-18

Empowering farmers through mobile applications

ADHIGURU P.[1] AND LETHA, DEVI G.[2]

[1] Principal Scientist, Division of Agricultural Extension, ICAR, New Delhi

[2] Senior Scientist, ICAR-NIANP, Bengaluru

Since the Green Revolution in the 1960s, India's production of food grain has increased fivefold. The country is now self-sufficient, but it is facing a major issues. Differences in the pace and patterns of technological and institutional changes and environmental conditions have been identified as the crucial factors causing uneven patterns of agricultural growth. Indian agriculture facing challenges of diversified farming practices, resource generation and soil problem. For resource generation rural youths are migrating from rural to urban areas. In order to address the diverse needs over the period the scenario has changed a lot due to increasing demands of timely and accurate information. Due to increasing agricultural demands farmers need timely and accurate information. Due to several constraints the conventional model of agricultural extension was not effective as the coverage was less. It can also be difficult to ensure the quality and consistency of the information delivered. (Cole and Fernando 2016). The public agricultural extension system which is responsible for disseminating agricultural information to farmers has become less effective, more time consuming and unsuccessful in meeting the requirements of those involved in agricultural production and there is a gap between the extension agent and the farmers (Mruthunjaya and Adhiguru, 2005). Acquisition of information is costlier than dissemination of information. Mobile agricultural services are becoming more informative because it can increase the reach, while reducing the information acquisition cost. Mobile is one of the tool which is being used as a information source to the farmers.

Critical information needs of farmers

There is a problem of increasing demand of timely, accurate and diversified information. Mobile phone has expanded the scope of information dissemination by covering all agricultural related information. Farmers need information to improve their farming practices and these information needs may include the crop related information, decision about crop to be grown, seed varieties, fertilizer application, plant protection measures, harvesting, post harvest, mandi and market prices. The information needs of farmers can be broadly grouped as indicated in below (Table 1).

Table 1. Farmers' Information Needs

Category	Examples	Typical Information Needs
Know-how	Crop choice Seed variety	What are options for new crops or seed varieties? Are there higher value crops or better seed varieties I could be planting?
Context	Weather Plant protection Cultivation best practice	When should sow? When should I harvest? Given my climate/soil What are cultivation best practices for my crops and soil? What inputs should use? How to best apply them? Where can I find them?
Market information	Market Prices Market demand Logistics	What are prices and demand in relevant markets? Has there been a transport breakdown?

In rural areas, people access to various channels for getting information. People tend to attach more importance for some channels compared to some other channels. This kind of behavior is due to credibility with which the audience see the communication source. More the people accessing a source, more the credibility can be assigned. For instance, there are many sources/channels of information through which farmers seek information on technologies. It has been noted that some farmers are exposed to a relatively larger number of information sources, and with greater frequency, than others. The 59th NSSO survey results show that other farmers, input dealers, mass media and extension, agents are some important sources generally utilized by farmers (Fig. 1).

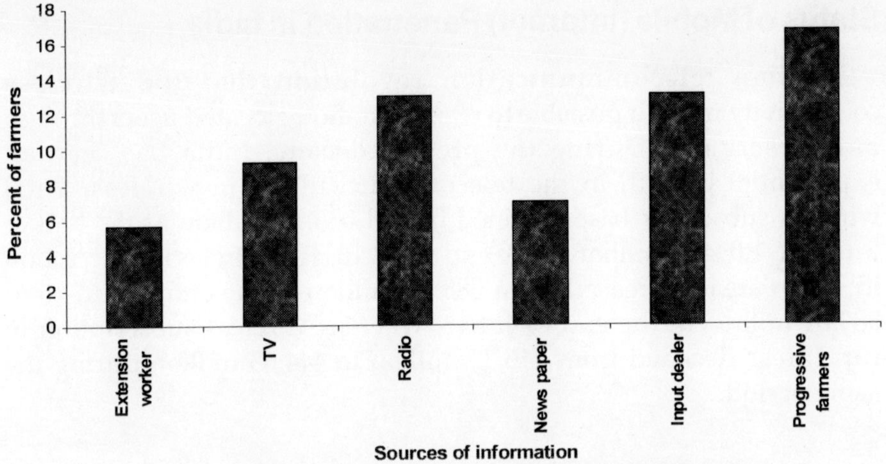

Fig. 1. Farmer Households Accessing Information

Table 2. Percentages of farmer households accessing modern agricultural technology

Source	% of households
Any source	40.4
Other progressive farmers	16.7
Input dealer	13.1
Radio	13
Television	9.3
Newspaper	7
Extension Worker	5.7
Primary cooperative society	3.6
Output buyers/food processor	2.3
Government demonstration	2
Village fair	2
Credit agency	1.8
Others	1.7
Participation in training	0.9
Krishi Vigyan Kendra	0.7
Para-technician/private Agency/NGO	0.6
Farmers' study tour	0.2

Source: NSSO (2005) Access to Modern Technology for Farming,Press Information Bureau, Government of India, New Delhi. http://www.mospi.nic.in/nsso_press_note_22june05.htm

Status of Mobile (internet) Penetration in India

Indian telecommunication revolution that too wireless connectivity made it possible to reach to remotly located users through mobile services. During the present decade, India has seen an exponential growth in the telecom particularly in wireless. Total wireless subscriber base declined from 1,035.12 million at the end of Jun-16 to 1,034.23 million at the end of Jul-16. The Wireless subscription in urban areas increased from 588.78 million at the end of Jun-16 to 589.61 million at the end of Jul-16, whereas wireless subscription in rural areas declined from 446.33 million to 444.63 million during the same period.

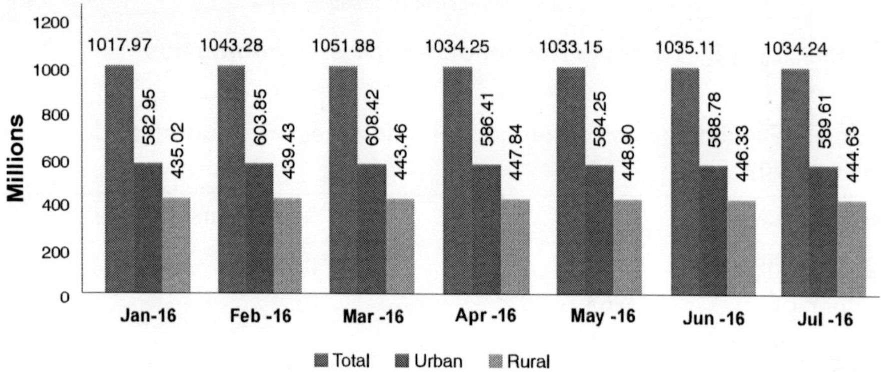

Fig 2. Wireless subscribers in India

Wireless Tele-density in India

The Wireless Tele-density in India declined from 81.26 at the end of Jun-16 to 81.11 at the end of Jul-16. The Urban Wireless Tele-density slightly declined from 148.03 at the end of Jun-16 to 148.00 at the end of Jul-16, and Rural Wireless Tele-density also declined from 50.95 to 50.72 during the same period. The share of urban and rural wireless subscribers in total number of wireless subscribers was 57.01% and 42.99% respectively at the end of Jul-16. (*Source: TRAI, 2016*)

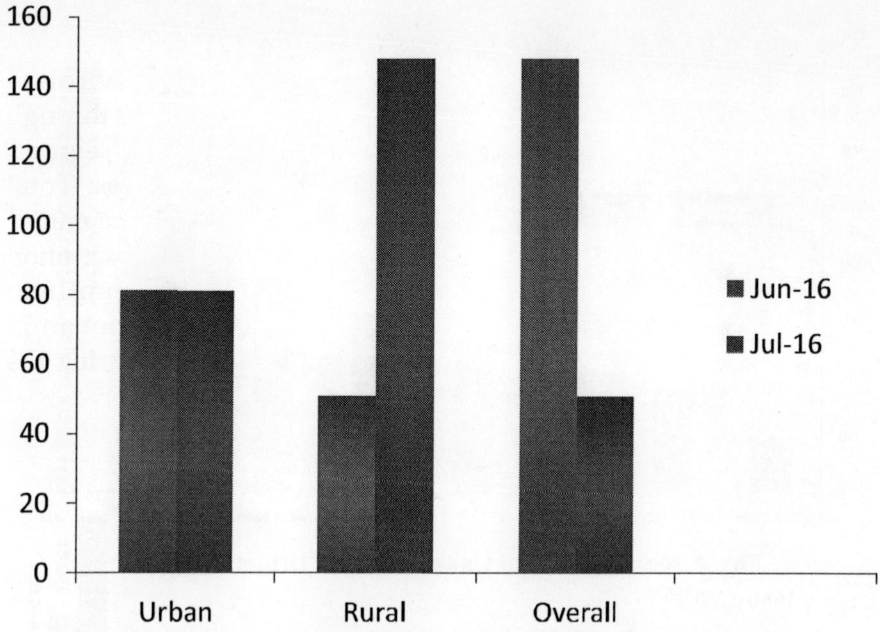

Fig. 3. Wireless tele-density in India

(*Source:* http://www.thehindu.com/news/cities/mumbai/business/with-220mn-users-india-is-now-worlds-secondbiggest-smartphone-market/article8186543.ece)

Internet penetration in urban and rural India

No. of mobiles, No. of internet connections

India is the 3rd largest in the world in terms of number of internet users next only to China and the United States of America.According to IAMAI 209 million urban Indian use internet.In urban areas 62 per cent users are male and 38 percent users are female. (Times of India, Dec, 4, pp.13) As per *Internet and Mobile Association of India (IAMAI)*, though there are about 109 million rural Indians are using internet. In rural areas 88 per cent users are male and 12 per cent users are females. IAMAI reveals internet penetration among rural females is growing at 61% and 79% among males unlike urban centres, internet use in rural areas is largely synonymous with mobiles and not so much with laptops, personal computers and other devices. India has become the second-biggest smartphone market in terms of active unique smartphone users, crossing 220 million users,Though India still has a long way to go as the smartphone penetration of the total potential population is still below 30 per cent.

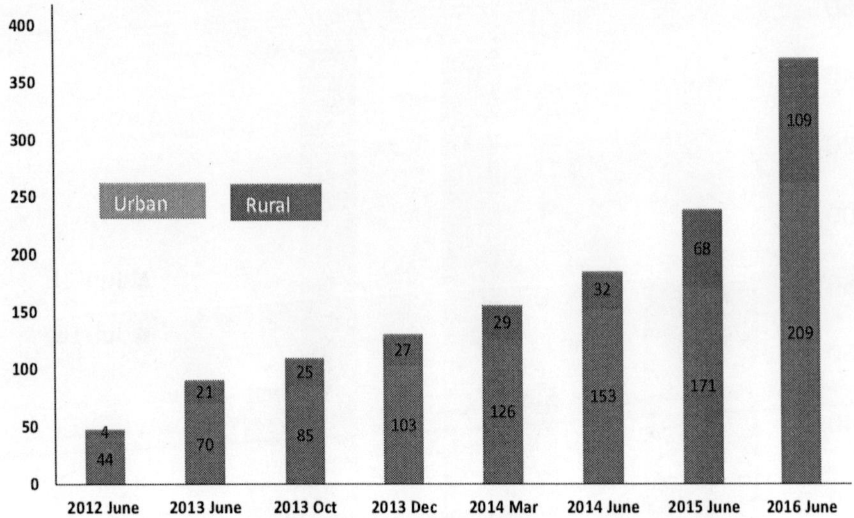

Fig. 4. Mobile Internet Users in India (Urban Vs Rural)

Source: IAMAI, 2016

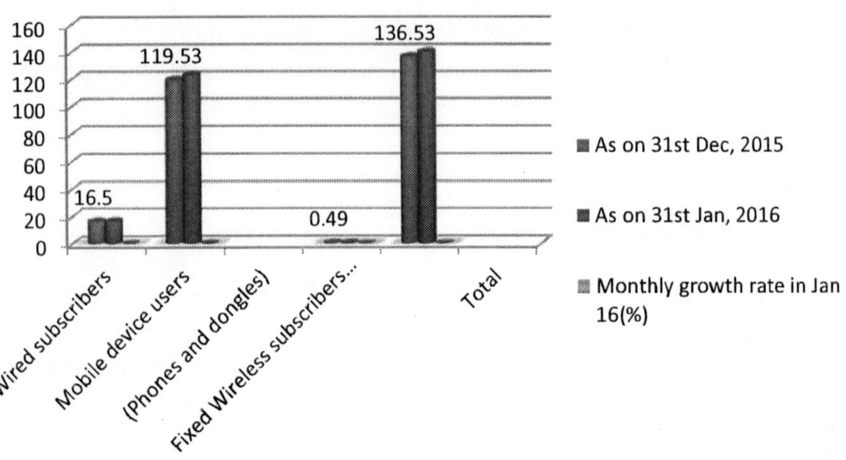

Fig. 5. Broadband subscribers (in million)

Source: IAMAI, 2016.

Usefulness of Mobile application to farmers

Type of information/services received through mobile is diverse and helps in cost-effective acquisition of information. In general, the advantages of this technology are:

1. The inevitable and growing importance of mobility in agriculture offers various advantages. All types of information on crop, soil,

climate, rainfall, seeds and machinery at any point in time, and any number of times is available on finger tips of farmers (World Bank 2011).

2. It gives a variety of information in single stroke. Mobile phones reduces the costs of farmers by reducing transaction cost, communication cost by providing them information at their field and as per demand.

3. The market connectivity is also improved with the visibility and knowledge of the potential buyers and sellers in the locality with an opportunity to develop direct contacts. The commodity prices can be delivered in a real time mode. (Kuek et al. 2011).

4. In addition to voice communication, mobile phones allow for the transfer of data, which can be used in the context of applications for the purposes of health, education, commerce or governance. (Rashid and Elder, 2009). After demonetisation usefulness of mobile phones has increased. Mobile services can enable rural community to receive and make payments and access financial services.

Innovative Initiatives

There are various innovative initiatives which are being provided by different organisations in agricultural information / services to the farmers by mobile application.The following mobile applications have developed by different institutions to provide agricultural information/ services to the farmers.

Name of the app	Name of the organization	Information/Service provided
Kisansuvidha app	ICAR-New Delhi	Weather, market prices, seeds, fertilizers, machinery etc.
Mobile app riceXpert	ICAR-National Rice Research Institute, Cuttack	Insect pests, nutrients, weeds, nematodes and disease-related problems, rice varieties for different ecologies, farm implements for different field and post harvest operations.
Pusa krishi app	IARI-New Delhi	New varieties of crops developed by Indian Council of Agriculture Research (ICAR), resource conserving cultivation practices as well as farm machinery and its implementation
Oilseeds at a glance	ICAR-Indian Institute of oilseeds research	Comprehensive information about 9 oilseed crops viz., Groundnut, Rapeseed and Mustard, Soybean, Sunflower, Safflower, Sesame, Niger, Castor and Linseed. Crop information, Released varieties and hybrids, State-wise preferred varieties and hybrids, AICRP centres contact information, ICAR seed Project centres dealing with oilseed crops, Mandis/ selling markets for each state.
Cotton mobile app	CICR Nagpur	Cotton cultivation, planning for cultivation, selection of BT varieties, cotton sowing, seed treatment, nutrient management, weed and water management, cotton growth management, pest management, picking and storage
KVK portal	ICAR-New Delhi	Information and advisory to the farmers and facilitate online monitoring of the KVK activities
mKisan SMS portal	Department of Agriculture & Cooperation	Enables all Central and State government organizations in agriculture and allied sectors to give

		information/services/advisories to farmers by SMS in their language, preference of agricultural practices and location.
IFFCOkisaan app	IFFCO	Information related to mandi prices, weather forecast, agricultural advisory, best practices tips related to agriculture, Animal Husbandry, horticulture; a buyer and seller platform, and all agriculture related news and govt. schemes.
mkrishi	TCS	Information on microclimate, local market price,enables farmers to send queries specific to their land and crop to receive personalized replies from agricultural experts, on their phones.
Digital Mandi	BSNL	Allow farmers to call up a (BSNL) number and get the rate of commodities, like vegetables, grains, pulses or fruits in any mandi (market) across the state
Agri market	Brinvik	This app is for agriculture traders and farmers in india. Farmers can see the commodity prices on their mobile anytime. App is covering most of the farm product from all over the indian agriculture markets, market yard.

Success Stories/ Case Studies

1. RUDI Sandesha Vyavhar (RSV)

Organization: Self-Employed Women's Association (SEWA), Vodafone Foundation, Cherie Blair Foundation for Women

Media used: Mobile phone

Services: RUDI is an agricultural cooperative founded on a simple model of procuring raw agricultural produce from marginal farmers at market price, adding value by cleaning, processing and packaging it, before selling RUDI-branded products at affordable prices through a network of rural saleswomen. RSV has been developed specifically for RUDI, by working directly with the women entrepreneurs in the network to understand the information required to successfully place an order and manage their own business.

Beneficiaries: Women

Mode of content: Gujarati

Impact: SEWA operates in 14 states of India building the capacities of over 1.9 million informal economy women workers. 1,300 illiterate and semi-literate women from remote rural & desert prone villages have been trained to use the application and the project is on track to reach 2,500 by May 2015. To date over RSV has increased women's business options, household nutrition and food security. The application won 'Best Mobile Product in Emerging Markets' at the 2014 Global Mobile Awards.

http://designexpo.businessfightspoverty.org/showcase/rudi-sandesha-vyavhar-rsv/

2. CaféMóvel: Mobile extension service for coffee growers in India

Organization: The mobile service is designed, deployed and maintained by CABI, with help from the Coffee Board of India , the International Coffee Organization and Common Fund for Commodities.

Media used: Mobile phone

Services: CaféMóvel - a mobile-enabled extension service that uses CABI's Direct2Farm database to give advisory support, the latest research and real-time information on market data, auction prices and weather available to coffee farmers.

Beneficiaries: Direct2Farm database to provide information and advisory support to around 150,000 coffee farmers in southern India.

Mode of content: In their local languages

Impact: Mobile ICT initiatives such as Café Móvel, bridges the gaps between labs to farms. And this was supported by various coffee growers' associations also expressed their interest in this new mobile based extension service and shown their willingness to use it for their benefit. http://www.cabi.org/projects/project/32810

3. mKisan – A mobile-based agricultural advisory service

Organization: The project, delivered jointly by CABI, Handygo Technologies, International Livestock Research Institute and Digital Green Donor GSMA

Media used: Feature phone with video capability

Services: Subscribers receive real-time and interactive advice direct from a panel of experts on crop and livestock such as insects, diseases and nutrition helpline. Farmers also receive regular weather bulletins, pest and disease alerts and market price information to support on-farm decision-making. Advice is delivered through voice based messages using IVR (interactive voice response) technology available on all mobile phone types. This has the advantage of making the service widely accessible, while overcoming lower literacy among the rural poor.

Beneficiaries: Farmers

Mode of content: Multi language

Impact: Since the service was launched in mid-2012, one million farmers have used the service. Around 340,000 have on-going subscriptions although peak months of activity occur at the start and end of the season when cropping decisions and market price information are of prime importance. over a million subscribers across six states of India; Andhra Pradesh, Bihar, Madhya Pradesh, Maharasthra, Karnataka and Uttar Pradesh.

4. Reuters Market Light

Organisation: Reuters Market Light

Media used: Mobile phone

Sevices: RML subscribers receive information via text messages on mobile phones, which is personalized based on the stage of the crop cycle and at the *taluka* level for weather forecasts. They also provide information about local market prices and local and international commodity information.

Beneficiaries: Farmers

Mode of content: Users can receive information in one of the nine languages.

Impact: This initiative has recorded 1 million unique subscribers so far in an estimated 50,000 villages across 13 states.

Impact of Mobile Application in Agriculture

1. **Achieve faster productivity**: Agricultural productivity varies dramatically around the world. Many factors may contribute to differences in productivity of the crop including lack of access to credit to invest in more productive tools and techniques, lack of marketing facilities, insurance to manage weather risk, and small and scattered land holdings. Another possibility is that farmers lack information about how to increase their crop productivity. It can also be difficult to ensure the quality and consistency of the information delivered. One alternative is to deliver agricultural information to farmers via low-cost information and communications technologies like mobile phones. Mobile phones significantly have reduced communication and information costs for the rural people and incresesd the productivity of crops. (Aker, 2011).

2. **Improving market efficiency**: Mobile phones have been spreading fast among farmers and they are exchanging their marketing formation among each other. Farmers directly contact markets brokers and near cities for sell their product. Similarly farmers focus up-to-date market information from so-cial and business networks (Ilahiane, 2007).

Constraints of Mobile Application

The rural community, especially farmers, are facing many problems in the use of ICT tools such as mobile phones.Constraints in any system adversely affect the process, thus know how of various constraints associated with mobile application is equally important to technology developer. There are many constraints indicated such as knowledge, content and accessibility.

1. **Knowledge related constraints:** Farmers are often unaware to access tools and associated information which could potentially be due to lack of awareness of mobile app as personal constraints. Sometimes in rural areas, people keep mobile phones but they do not know availability of apps, the manner in which app is to be utilized; therefore, the such applications developed by institutes to provide agricultural information and services to the

farmers remain inadequately explored and exploited. In majority of cases, users are seen to use mobile phones for social purposes only. The illiteracy is also a cause of use mobile phone among farmers because famers could not contact with related officers and department and get information about different agricultural information.

2. **Content related constraints:** Main challenges that rural communities have faced is impeded mobile application due to language barrier and. Other problems being seen is difficulty in accessing accurate, timely and reliable information. Non-availability of content in vernacular languages as well as lack of pictorial interface in retrieving the content is biggest challenge for them.

3. **Accessibility**:Technical inefficiency in operating mobile phones is a barrier between the user and flow of information. Most of the farmers do not use Smartphone, so they cannot get information that is developed based on smartphone software,therefore, there is need to develop applications which can also provide information through SMS. A major constraint in accessibility is slow internet speed in rural India.

4. **Connectivity:** Farmers face low network coverage. In rural India due to poor connectivity and poor network coverage mobile phone application is slightly decreasing in recent years.

Strategies for Enhanced use of Mobile Application

1. Government and institutions can create awareness programme on potential of mobile phones and its usage.

2. The content and design should be user centric. Institutions which are being developing apps can provide accurate, pictorial and provide content in their local languages so that farmers can easily access the information. Smart phone should be designed in such a way that interface for farmers especially both illiterate and literate can use it properly.

3. Imparting technical training to farmers on use of Apps and retrieval of information will play most important role in extending the reach of mobile application for empowerment of farmers. So that they can utilize these technical training for operating mobile phone.

Summary

Mobile application has a great potential for delivering agricultural information and services. The mobile application is a revolution in ICT and is catering to information related needs of farmers. Various institutions and organizations are developing different apps to disseminate accurate and timely information to enable farmers with knowledge of farming activities.There is significant range of benefits arising from the use of mobiles in the context of rural people. However, some sections of farmers are facing problems in utilizing information delivered through mobile phones. There is a need to identify suitable strategies for enhanced use of mobile application in order to enhance use of scientific information for the benefit of farmers and eventually our nation.

Reference

Aker, J.C. (2011). Dial "A" for agriculture: a review of information and communication technologies for agricultural extension in developingcountries. *Agricultural Economics*. 42 (6): 631-647.

Chitra *et al.* (2016). Smartphones are helping rural women get better at business.*Times of India*. Dec, 4, pp.13.

Cole *et al.* (2016). " 'Mobiliziing Agricultural Advice: Technology Adoption, Diffusion and Sustainability." *Harvard Business School Working Paper*. Cambridge. April, 2016.

lahiane, H. (2007). Impacts of Information and Communication Technologies in Agriculture:Farmers and Mobile Phones in Morocco." Paper presented at the Annual Meetings of the American Anthropological Association. December 1, Washington DC.

M. Kirk *et al.* (2011). Connected Agriculture: The role of mobile in driving efficiency and sustainability in the food and agriculture value chain. Vodafone and Accenture, Report 2011.

Mruthunjaya H and Adhiguru A (2005). ICT for livelihood security: a reality check. *Mainstreaming ICTs.* 2(2):14-18.

Oiang *et al.* (2011). Mobile Applications for Agriculture and Rural Development. Report of world bank.

Rashid A.T., And Elder, L. (2009). Mobile Phones and Development: An Analysis of IDARC-Supported Projects. *The Electronic Journal on Information Systems in Developing Countries.* 36(2): 1-16.

http://mkisan.gov.in/aboutmkisan.aspx

http://designexpo.businessfightspoverty.org/showcase/rudi-sandesha-vyavhar-rsv/

http://timesofindia.indiatimes.com/city/hyderabad/Bhoochetana-A-major-agricultural-success-story-in-India-in-decades/articleshow/49871275.cms

World Bank, InfoDev. (2011), ICT in Agriculture Sourcebook, Agriculture and Rural Development.

http://www.thehindu.com/news/cities/mumbai/business/with-220mn-users-india-is-now-worlds-secondbiggest-smartphone-market/article8186543.ece

□□□

Chapter-19

Doubling farmers' income by 2022 through group approach: Strategies and policy perspectives

DEBESH ROY

DGM, Department of Economic Analysis and Research, NABARD, Mumbai

Agriculture provides employment to about 48.9 percent of total workforce in India, but contributes only 15 percent to country's Gross Value Added (GVA)[1]. About 85 percent of operational holdings in country is in small and marginal categories, and average size of an operational holding is only 1.15 ha. Due to fragmentation and disorganisation, farmers face constraints in procuring inputs like seeds and fertilizers at reasonable prices, lack bargaining power in market for realising better value for their produce, and have inadequate access to credit, technology and extension services. As a result, majority of smallholders are unable to realise optimal value from their farming operations.

Stagnation in improvement in productivity in agricultural crops has been observed in India, due to following factors:

i) Decline in farm size and income

ii) Depleting natural resource base

iii) Increasing input costs, and adverse economics of farming

iv) Deficiency of micronutrients in soil and deteriorating soil health

v) Inadequate post-harvest technology

vi) Uncertain market prospects; and

vii) High indebtedness of farmers.

---

[1] Second Advance Estimates of National Income 2016-17, Ministry of Statistics and Programme Implementation, Government of India, 28 February 2017.

Government of India has envisioned achievement of doubling of farmers' income by year 2022. Following seven-point strategy for doubling farmers' income announced by Hon'ble Prime Minister needs to be adopted by central and state governments, and other stakeholders, viz. RBI, NABARD, banks, insurance companies, developmental agencies, research institutions/agricultural universities, extension agencies, corporate sector, cooperatives, producer organisations/producer companies, etc., for achieving target within stipulated time:

i) Irrigation with focus on water-use efficiency "per drop more crop"
ii) Quality seed and soil health
iii) Investments in warehouses and cold chains
iv) Value addition through food processing
v) Electronic National Agriculture Market (e-NAM)
vi) Increase in coverage and effective implementation of new crop insurance scheme–Pradhan Mantri Krishi Bima Yojana (PMKBY)
vii) Promotion of ancillary activities like dairy, poultry, bee-keeping and fisheries. Adoption of this strategy would require massive investments in irrigation through Pradhan Mantri Krishi Sinchai Yojana (PMKSY), appropriate technology, agricultural research and extension, agricultural market infrastructure, and agriculture value chains, supported by adequate and timely availability of bank credit and crop insurance, leading to a significant increase in agricultural production and productivity. Diversification in form of product (high value enterprises), process (precision farming), and time (delinking from seasonality), could be a major game changer for achieving doubling of farmers' income (NABARD, 2016).

A group approach could help small and marginal farmers to undertake lumpy investments by pooling financial resources (Agarwal, 2016), especially by availing bank credit. It could also enable diversification of activities, and aggregation of produce for marketing. This paper examines strategies and policy support for doubling farmers' income by 2022 through a group approach, with a special focus on role and effectiveness of farmer producer organisations (FPOs).

Status of Small and Marginal Farm Households in India

Situation Assessment Survey of Agricultural Households in India (NSS 70th Round)[2] displays average monthly income from different sources, total consumption expenditure, net income, and net investment in productive assets per agricultural household during agricultural year July 2012-June 2013 for each size class of land possessed. It was observed that small land holders were mostly dependent on wage/salary employment than farm business (cultivation and farming of animals) for their income. For households belonging to lowest size class, rearing of animals fetched more income than cultivation. Percentage share of income from cultivation/farm business in average monthly income increased with increase in land possession. Share of income from non-farm business in average monthly income decreased with increase in land size. Similarly net investment in productive assets per agricultural household increased with increase in land size. Further, net monthly income (farm and non-farm) in respect of size classes up to 1 ha was negative. However, net income increased steadily with increase in size class.

Average monthly income per agricultural households during agricultural year July 2012- June 2013 was estimated as ₹ 6,426, while net income stood at a minuscule ₹ 203 (Table 1). Net receipt from farm business (cultivation and farming of animals) accounted for about 60 percent of average monthly income per agricultural household during agricultural year July 2012 - June 2013. More than 32 percent of income was contributed by wage/salary employment (Table 1). average net monthly income in respect of farm households in categories of landless (<0.01 ha), lower marginal (0.01-0.40 ha) and upper marginal (0.41-1.00 ha) was observed to be negative at ₹ (-) 547, ₹ (-) 1,249 and ₹ (-) 773, respectively. In case of small agricultural households (1.01-2.00 ha), average net monthly income was a meagre ₹ 891 (Table 1). Average monthly expenditure incurred on productive assets used in farm business showed a wide variation across size classes of land possessed, ranging from less than one-third (₹ 281) of all India average (₹ 1,013) in lowest size class of land possessed, to more than nine times all India average in case of highest size class (₹ 9,568) (Table 1).

---

[2] *Income, Expenditure, Productive Assets and Indebtedness of Agricultural Households in India*, NSS 70th Round Report 576 (70/33/3), Ministry of Statistics and Programme Implementation, GoI, 2016.

Table 1. Average monthly income from different sources and consumption expenditure per agricultural household (July 2012-June 2013) for each size class of land possessed

Size class of land (ha)	Average monthly income (₹)					Average monthly consumption expenditure (₹)	Net monthly income (₹)	Average monthly expenditure on productive assets in farm business (₹)
	Income from wages/ salary	Net receipt from agricultural activities		Net receipt from non-farm business	Total income			
		Cultivation	Rearingof animals					
Landless (<0.01)	2902 (63.6)	30 (0.7)	1181 (25.9)	447 (9.8)	4561	5108	-547	281
Lower marginal (0.01-0.40)	2386 (57.5)	687 (16.5)	621 (15.0)	459 (11.0)	4152	5401	-1249	287
Upper marginal (0.41-1.00)	2011 (38.3)	2145 (40.9)	629 (12.0)	462 (8.8)	5247	6020	-773	837
Small (1.01-2.00)	1728 (23.5)	4209 (57.3)	818 (11.1)	593 (8.1)	7348	6457	891	1741
Semi-medium (2.01-4.00)	1657 (15.4)	7359 (68.6)	1161 (10.8)	554 (5.2)	10730	7786	2944	1667
Medium (4.01-10.00)	2031 (10.3)	15243 (77.6)	1501 (7.6)	861 (4.4)	19637	10104	9533	2805
Large (>10.00)	1311 (3.2)	35685 (86.2)	2622 (6.3)	1770 (4.3)	41388	14447	26941	9568
All sizes	2071 (32.2)	3081 (47.9)	763 (11.9)	512 (8.0)	6426	6223	203	1013

Note: Figures in brackets indicate percentage share of total income.

Source: Statements 3.2 and 3.7 of Income, Expenditure, Productive Assets and Indebtedness of Agricultural Households in India, NSS 70th Round Report 576 (70/33/3), Ministry of Statistics and Programme Implementation, GoI, 2016.

Studies have shown that small-scale farming is inefficient in India (e.g. Foster and Rosenzweig, 2010). Lives of smallholding families can be improved only by building on their higher per acre agricultural productivity and by promoting off-farm rural employment (Chand *et al.*, 2011). Therefore, strategies for promoting agricultural growth and sustainability, and significantly increasing income of small and marginal farmer households, should include a group approach in form of FPOs, producer companies, cooperatives or collectives, diversification to high-value agriculture, allied activities (especially dairy, poultry, fisheries, bee-keeping, etc.) and promoting off-farm work in rural areas.

A group approach can increase market access to land, and help spread risk of farming among a larger number and increase production opportunities by experimenting with higher value, more risk prone crops with larger payoffs (Agarwal, 2016). It also allows labour sharing and easy substitution. Further, a group would be better placed to enter into non-exploitative contract farming arrangements (ibid).

Group Approach to Smallholder Agriculture – Farmer Producer Organisations

1. Farmer Producer Organisations – Role of Government of India

By organising themselves collectively, producers can utilize scale to procure inputs at a lower price, and gain more selling power for their produce/product. Farmer producer organisations (FPOs) can provide a strong platform to take up collective business activities, build capacity and provide linkages to markets. FPOs, registered under special provisions of Companies Act, 2013[3], are most appropriate institutions for mobilising farmers and building their capacity to collectively leverage their production and marketing strength. Government of India has taken various policy initiatives, viz. dedicated source of funding to FPOs from Rashtriya Krishi Vikas Yojana (RKVY), Equity Grant and Credit Guarantee Fund Scheme for FPOs by Small Farmers Agribusiness Consortium (SFAC), and setting up of Producers Organizations Development Fund (PODF) and Producers' Organisation Development and Upliftment Corpus (PRODUCE) Fund in NABARD.

[3] Reference: Section 465(1) of Companies Act, 2013 on Producer Companies (Part IXA of Companies Act, 1956)

SFAC provides support to equity base of farmer producer companies (FPCs), by providing matching equity grants, up to a maximum of ₹ 10 lakh; and setting up of a Credit Guarantee Fund to provide risk cover to banks which provide loans to FPCs without collateral up to ₹ 100 lakh (SFAC, 2016). SFAC has also been mandated by Ministry of Agriculture and Farmers Welfare, GoI to support state governments in formation of FPOs.

2. NABARD's Initiatives for Farmer Producer Organisations

PODF set up in NABARD in 2011–12, supports FPOs to spearhead promotional, credit-plus activities, aggregation, processing and marketing activities to ensure better income realisation to its members. Major activities of FPOs assisted by NABARD so far, include establishment of bamboo processing units, dairy activities, agro-service centres, agro-tourism centre, green tea manufacturing factory, spices processing unit and bio-processed and commercial production of fish.

PRODUCE Fund with a corpus of Rs. 200 crore for promotion of 2,000 FPOs by 31st March 2016, was a major intervention aimed at facilitating aggregation of input supplies and output marketing, besides village level value addition. As on 31 March 2016, 2,173 FPOs were formed under Fund, in various parts of country (NABARD, 2016a). These FPOs had been handheld with a larger concentration (more than 100 FPOs per state) in states like Karnataka, Rajasthan, Tamil Nadu, West Bengal, Madhya Pradesh, Uttar Pradesh, Odisha, Maharashtra, Kerala, Gujarat Bihar, and Andhra Pradesh. A grant assistance of ₹ 125.97 crore was sanctioned during 2015–16 for promotion and capacity-building/nurturing of these FPOs by Producer Organisation Promoting Institutions (POPIs) concerned (ibid.). A National Advisory Committee (NAC) was constituted under chairmanship of Chairman, NABARD, with representatives from GoI, SFAC, corporates, academicians, bankers and practitioners in field. NABARD had partnered with 785 POPIs and 16 Resource Support Agencies (RSAs) in various states as on 31st March 2016, for providing support for nurturing and hand holding of FPOs over a period of 3 years (NABARD, 2016a).

NABKISAN Finance Limited (NKFL), a subsidiary of NABARD, provides support for livelihood/income generating activities by extending credit to Panchayat Level Federations, trusts, societies and companies/MFIs for on-lending to its member SHGs/JLGs. It also supports FPOs for term loan and working capital requirements.

Group Approach for doubling Farmers' Income by 2022: Strategies and Policy Perspectives

1. Water-use Efficiency-Per Drop More Crop

Inefficiency in use of water has led to depletion in ground water availability and low agricultural productivity in country. Incidentally, India uses 2 to 4 times more water to produce a unit of major crops than does China and Brazil (GoI, 2016 a). In order to raise agricultural productivity in India, it is imperative to expand acreage under irrigation along with adoption of appropriate technologies for efficient utilization of water through suitable pricing (GoI, 2016).

Net irrigated area to total cropped area in country stood at 33.9 percent (2012-13), and there is regional disparity in irrigated farming, with net irrigated area to total cropped area at more than 50 percent in states of Punjab, Tamil Nadu and Uttar Pradesh, while it is less than 50 percent in remaining states (GoI, 2016). Therefore, policy focus should be on creating and utilizing greater irrigation potential, reducing regional disparities in irrigation potential, and improving irrigation efficiency.

Water productivity at all India level is very low and needs to be enhanced through tapping, harvesting and recycling water, efficient on-farm water management practices, minor irrigation, and use of waste water and resource conservation technologies. Overall irrigation efficiency of major and medium irrigation projects in India is estimated at around 38 percent. Efficiency of surface irrigation system can be improved from about 35-40 percent to around 60 percent and that of groundwater from about 65-70 percent to 75 percent (GoI, 2016). GoI has formulated PMKSY with vision of extending coverage of irrigation *'har khet ko pani'* (water to every farm) and improving water use efficiency, viz. 'per drop more crop' in a focused manner with end to end solution on source creation, distribution, management, field application, and extension activities. Scheme involves convergence of investments in irrigation at field level, expansion of cultivable area under assured irrigation, improvement in on-farm water use efficiency to reduce wastage of water, enhancement of adoption of precision-irrigation and other water saving technologies, enhancement in recharge of aquifers and introduction of sustainable water conservation practices. A dedicated Long Term Irrigation Fund (LTIF) in NABARD with an initial corpus of ₹ 20,000 crore was instituted by

GoI in NABARD, and another ₹ 20,000 was announced in Union Budget 2017-18, for fast-tracking implementation of incomplete major and medium irrigation projects under PMKSY.

While micro-irrigation techniques, owing to high fixed costs of adoption, have mostly been used for high value crops, recent research has shown its feasibility even in wheat and rice (GoI, 2016 a). Union Budget 2017-18 announced setting up of a dedicated Micro Irrigation Fund in NABARD, with an initial corpus of ₹ 5,000 crore to achieve goal of 'per drop more crop'. FPOs, with their pooled resources, could be a cost-effective option for adoption of micro-irrigation techniques for cultivation of wheat and rice, apart from high value crops.

2. Seed, Fertilizers and Pesticides

There are multiple challenges to development and adoption of quality seeds in India, viz. high cost of seeds for small and marginal farmers, inadequate supply of quality seeds, non-resolution of issues related to adoption of genetically modified (GM) crops and lack of competition in seed market. Green revolution was driven by indigenously developed high yielding varieties (HYV) seeds. Therefore, in order to achieve a second green revolution there is an urgent need to develop seeds/seed technology in both public and private sectors. Under circumstances, FPOs could be incentivised to play an important role in developing seeds and seed technology in a cost-effective manner.

Fertilizer use in our country is inefficient. Imbalances in use of fertilizers are manifested by excessive dependence on urea owing to low/distorted prices of urea. Further, there has been diversion of subsidised fertilizers to non-agricultural use. While indiscriminate use of fertilizers has not improved yield of crops proportionately, it has resulted in depletion of soil fertility in many areas. Thus, there is a need to reform present fertilizer subsidy regime, by moving towards an input, crop and region neutral format and also to minimize diversions. Linking soil health cards to provide profile of soil and fertilizer on basis of same profile utilizing fertilizer, even if not subsidized, can improve yield of crops (GoI,2016). Efficient nutrient management for improved productivity is a huge challenge, which can be addressed through information technology and providing soil fertility maps to FPOs.

Farmer's crop yield losses in country range from 15 to 25 percent due to presence of weeds, pests, diseases and rodents. However, per hectare pesticide use in India is very low. India uses 0.5 kg per ha pesticide compared to 7.0 kg per ha in USA, 2.5 kg per ha in Europe, 12 kg per ha in Japan and 6.6 kg per ha in Korea. There is a need for farmers to be educated about classification of insecticides on basis of their toxicity. There is also a need for greater focus on Integrated Pest Management (IPM). Further, being environment friendly, non-toxic and cost effective, bio-pesticides need to be promoted among small farmers to improve productivity in agriculture (GoI, 2016). This could be done effectively through FPOs.

3. Research and Development and Agricultural Extension

Research strategy should be pro-nature and small farmer-oriented (Swaminathan, 2010). There is a need to shift focus of research and development agenda from crop-centric research in irrigated areas to location-specific cropping systems in dry lands, hills and tribal areas (*ibid.*). Greater attention needs to be directed towards horticulture crops, which are land and water-saving by nature. Efforts may also be made to harness remote sensing technologies to optimize application of inputs, and explore areas in emerging capital-intensive biotechnology (Swaminathan, 2010). Involving FPOs in whole process could lead to achievement of desired results.

Needs of farmers in terms of information and technology support have become more complex, due to rapid pace of developments in agricultural sector. Climate change, depleting natural resources, scarcity of labour and volatile market forces are some of concerns that have put tremendous pressure on both farmers as well as extension systems in context of increasing productivity, profitability and sustainability of agriculture.

Agricultural extension services have to be revitalized by making it more relevant, useful and timely in order to improve agricultural productivity (GoI, 2016). Extension services also need to be done in an input, crop and region neutral way, and should aim at minimizing wastage in inputs and in produce, till it leaves farm gate. Further, efforts need to be made to enhance post-harvest processing and value added activities at farm. FPOs, Farmers Clubs/Farmers Federations could be important agencies for providing extension services to farmers.

4. Capital Formation in Agriculture

Capital formation in agriculture is of critical importance for sustainability of agricultural growth. Percentage share of Gross Capital Formation (GCF) in agriculture and allied sector in Gross Value Added (GVA) from agriculture declined from 18.3 percent in 2011-12 to 12.6 percent in 2015-16 (Table 2). Considering importance of agriculture and allied sector in increasing income, employment and achieving inclusive growth, it is imperative to arrest declining trend and increase investment in sector.

Declining GCF to GVA ratio in agriculture can be attributed to decline in public sector investments. While share of public sector in GCF had declined from above 20 percent in 2004-05 to 16.8 percent in 2013-14, share of private sector increased from 78 percent to 83 percent during same period (GoI, 2016). In order to enable acceleration of private sector investments in agriculture, banks could consider financing FPOs/FPCs for warehouses, cold storages, food processing units, milk chilling plants, etc.

Table 2. GVA and GCF in Agriculture and Allied Sector (Amount in Rs. Cr)

Year	Agriculture and allied sector		Share of GCF in GVA of Agriculture & allied sector (%)
	GVA (at basic prices) (constant 2011-12 prices)	GCF (at constant 2011-12 prices)	
2011-12	1501816	274432	18.3
2012-13	1680797	253279	15.1
2013-14	1902452	284134	14.9
2014-15	1995251	277436	13.9
2015-16	2093081	263147	12.6

Source: Handbook of Statistics on Indian Economy 2015-16, RBI., and *Key Economic Indicators, February 2017, Office of Economic Adviser, Ministry of Commerce & Industry, GoI.*

Public sector investment in irrigation, rural roads, power, telecommunications, and market infrastructure, research, and extension services results in high growth of agricultural sector and reduction in poverty. Considering existence of fiscal constraints, state governments' investment strategy in agriculture should be guided by efficient and equitable use of resources with high payoffs.

Advantages of improved technology in farm mechanisation, that help to overcome productivity challenges in agriculture, often elude small and marginal farmers on account of lack of affordability and bulkiness of investment. Custom hiring centres, especially of FPOs could facilitate hiring of machines and implements suited to farm and crop requirements and enable timeliness and precision, leading to improvement of productivity per unit of crucial inputs. Channellising incentives (in form of subsidies) to FPOs for custom hiring centres, therefore, will generate more externalities, and improve net income of small farms.

5. Climate Smart Agriculture

Agriculture is most vulnerable sector to climate change. It would be a daunting challenge to produce enough food for increasing population in face of decreasing resources and changing climate. Changes in rainfall patterns can result in severe water shortages or floods. Rising temperatures can cause changes to crop growing seasons or even reduce yields of crops. There will also be a further knock on effects on environment which will impair agricultural production as a whole (FAO 2015).

It is felt that reactive adaptations like improving markets, changing agricultural policies, enhancing social protection and preparing for disasters have their limitations. More adaptiveimprovements are required instead. Therefore, concept of Climate-Smart Agriculture (CSA) was introduced, as an integrative approach to address interlinked challenges facing food security and climate change through three dimensions of sustainable development, viz.,

- Economic–sustainably increasing agricultural productivity, to support equitable increases in farm incomes, food security and development
- Social–adapting and building resilience of agricultural and food security systems to climate change at multiple levels; and
- Environmental–reducing and/or removing greenhouse gas (GHG) emissions from agriculture (including crops, livestock and fisheries) (FAO, 2015).

CSA entails greater investment in managing climate risks, understanding and planning for these adaptive changes that are required, and making use of opportunities for reducing or removing GHG emissions wherever possible. It relates to actions both on-farm and beyond, and includes technologies, policies, institutions and investments. Approaches include:

- Management of farms, crops, livestock, aquaculture and capture fisheries to manage resources better
- Ecosystem and landscape management to conserve ecosystem services that are critical for increasing resource efficiency and resilience; and
- Services for farmers and land managers to ensure they are able to implement necessary changes (FAO, 2015).

Impact of climate change on agriculture in India could result in problems with food security and may threaten livelihood activities upon which majority of population depends. CSA needs to be encouraged in India to address problem of agrarian distress, especially through FPOs, as it could lead to enhancement in farmers' incomes.

6. Efficient Agri-Value Chains

India is second largest producer of food in world after China. This sector is a major generator of employment with around 65 lakh people engaged in various food processing activities in formal sector. Strength of sector lies in fact that India is leading producer in a number of commodities like milk, cereals, tea, and sugarcane. It is also second largest producer of fruits and vegetables after China, and fifth in poultry production in world.

However, a major cause for concern lies in poor post-harvest management infrastructure facilities. Post-harvest wastage in country has been estimated to tune of ₹ 44,000 crore per annum. Wastage occurs at all levels of value chain at levels of farmer, transporter, wholesaler and retailer. This could be due to crop damage, improper harvesting techniques, poor packaging, poor transportation, poor handling, multiple handling, lack of storage infrastructure, grading, sorting, and moisture loss at various stages of value chain. Therefore investment in agri-value chain infrastructure needs to be stepped up.

India has a huge domestic market for food and food products. Indian consumers spend more than 30 percent of their income on food products (www.mofpi.nic.in). Food industry is currently growing at an average rate of 8.4 percent a year, way ahead of agriculture and manufacturing. Apart from this, India has locational advantage of being nearer to major markets of Middle East, South-East Asia and Africa. Given size of processed food market, there is enough space for more investment in food processing sector in India.

Policy focus on agri-value chains needs to be on promoting inclusive value chain, which would benefit small and marginal farmers, through value identification, value creation, and value capture but more importantly, from value sharing with smallholders (Singh, 2016). An effective way to make it possible is through efficient FPOs/ FPCs.

India is largest producer of milk in world, accounting for 17 percent of world's production. The per capita availability of milk in India increased from 176 grams per day in 1990-91 to 322 grams per day by 2014-15. It is more than world average of 294 grams per day during 2013 (GoI, 2016). This represents a sustained growth in availability of milk and milk products for growing population. Dairying is an important secondary source of income for millions of rural households engaged in agriculture. Success of dairy industry has resulted from integrated co-operative system of milk collection, transportation, processing and distribution, conversion of same to milk powder and products, to minimize seasonal impact on suppliers and buyers, retail distribution of milk and milk products, sharing of profits with farmer, which are ploughed back to enhance productivity and needs to be emulated by other farm produce/ producers (GoI, 2016). A focus on dairy value chain could significantly improve income of smallholder farmers. Therefore, in order to double income of farmers, it is important to focus on dairy value chain through FPOs/ FPCs.

India is second largest producer of fruits and vegetables in world. Productivity of horticultural crops increased by about 34 percent between 2004-05 and 2014-15[4]. Significantly, while area under horticulture grew by about 3.8 percent per annum, production rose by 7.6 percent per annum over last decade. In poultry segment, GoI's focus, besides framing suitable policies for enhancing commercial poultry production, is for strengthening family poultry system, which addresses livelihood issues. Further, fisheries constitute about one percent of GDP of country and 5.08 percent of agriculture GDP. Promotion of value chains for horticulture, poultry and fisheries,

[4] *State of Indian Agriculture 2015-16*, Ministry of Agriculture and Farmers Welfare, GoI.

involving FPOs/ FPCs, could significantly give a fillip to income of small and marginal farmers.

Promotion of off-farm sector (OFS), including MSME and services sectors, assumes significance in context of reducing over dependence of rural population on agriculture and to provide alternative livelihood options for farmers. These activities could be made profitable as part of food and agri-value chains.

7. Tenancy Reforms

A major problem confronted by small and marginal farmers in India is absence of proper secured land titles and predominance of tenant cultivators. There is a strong case for legalisation and liberalisation of land leasing as it would help promote agricultural efficiency, equity, occupational diversification and rapid rural transformation. According to Report of Expert Committee on Land Leasing (NITI Aayog, 2016), legalisation of land leasing would ensure security of land ownership right for land owners, which in turn would provide security of tenure to tenants; help improve tenant farmers' access to credit, insurance and input use and consequently productivity of leased in land; better utilization of available land and labour, and increased farm output; and help improve agricultural efficiency.

Committee has also prepared a model agricultural land-leasing act, which needs to be enacted by state legislatures. Legalisation of land leasing would encourage large land owners to lease out land without fear of losing their land ownership rights and invest in non-farm enterprises, which is vital for occupational diversification and rapid rural transformation. This will reduce pressure of population on agriculture and enable small farmers to augment their size of operational holdings by leasing in land (NITI Aayog, 2016). Further, enactments of land leasing laws could encourage FPOs to facilitate aggregation of land leased out to small and marginal farmers, thereby enabling increase in production and productivity, and processing and marketing of agricultural produce.

8. Agricultural Credit

Credit is a critical input for accelerating agricultural production and productivity. There has been an impressive growth in agricultural credit flow from ₹ 2.29 lakh crore to ₹ 8.77 lakh crore during 2006–07 to 2015–16 (Chart1). However, share of term loan in total agricultural credit disbursed declined steadily from 35.4 percent in 2006–07 to 19.6 percent in 2015–16 (Chart 2). There is an urgent need to increase share of term loan, which could lead to much-needed increase in capital formation in agriculture, and consequent increase in agricultural production and productivity (NABARD, 2016a). Agricultural credit is expected to cross ₹ 9 lakh crore during 2016-17, and target for 2017-18 has been fixed at Rs.10 lakh crore by Union Budget 2017-18. However, it must be emphasised that there is a need to increase share of investment credit to increase capital formation.

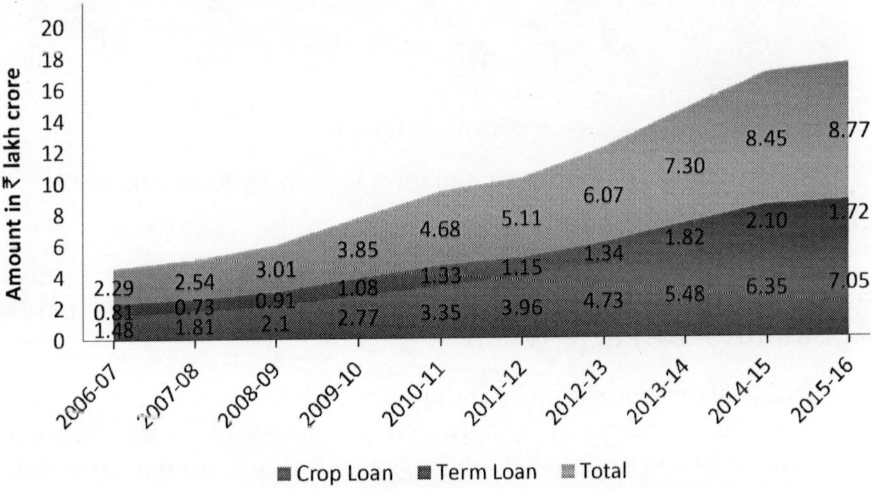

Chart 1. Credit flow to agriculture

Source: NABARD Annual Report 2015-16.

Therefore, banks need to think out-of-box by providing investment credit for agriculture and allied activities to FPOs. Group approach towards financing smallholders and landless labourers also includes financing Joint Liability Groups (JLGs). JLGs are meant for augmenting flow of credit to small/marginal, and tenant/landless farmers, extending collateral free loans to them and building mutual trust and confidence between banks and JLG members. During 2015–16, 5.93 lakh JLGs were promoted and financed by banks, taking cumulative number of JLGs promoted and financed by banks to 17.2 lakh as at end of March 2016. Apart from extending 100 percent refinance support to banks NABARD extends financial support for awareness creation and capacity building of all stakeholders of programme.

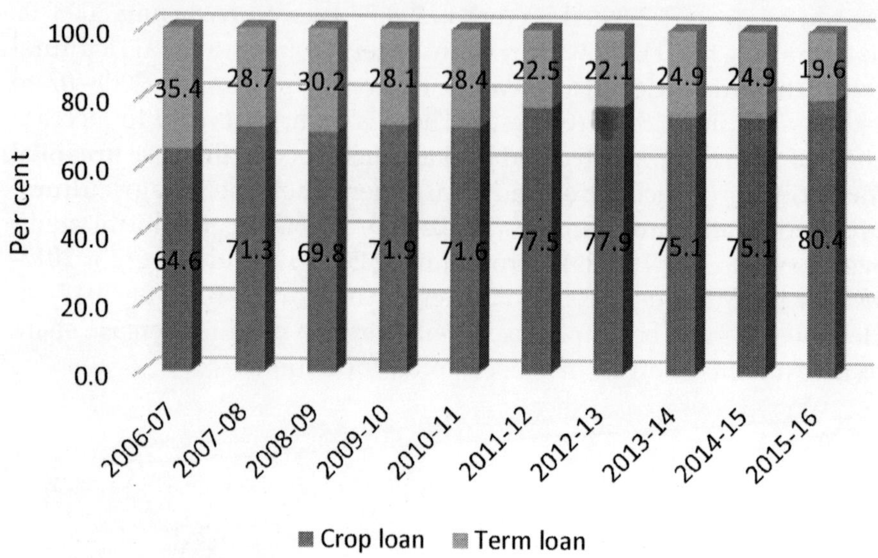

Chart 2. Share of crop loan and term loan in agricultural credit

Source: NABARD Annual Report 2015-16.

9. Agricultural Marketing Reforms – Electronic National Agriculture Market (e-NAM)

A critical problem faced by India's agricultural sector is fragmented and distortions-ridden state of agricultural markets. Electronic National Agriculture Market (e-NAM) envisages setting up of an appropriate common e-market platform that would be deployed in 585 regulated wholesale markets in States/Union Territories (UT) desirous of joining e-platform.

The e-NAM is envisaged as a win-win solution for all stakeholders. For farmers, e-NAM promises more options for sale at his nearest *mandi*. For local trader in *mandi*, e-NAM offers opportunity to access a larger national market for secondary trading. Bulk buyers, processors, exporters etc., benefit from being able to participate directly in trading at local *mandi* level through e-NAM platform, thereby reducing their intermediation costs. e-NAM could lead to significant benefits to farmers through higher returns, while benefitting buyers through lower transaction costs and consumers through stable prices. It will also facilitate emergence of integrated value chains in major agricultural commodities across country, and encourage setting up of scientific storage and movement of agri-commodities.

As at end of November 2016, 262 *mandis* across 11 states had gone live under e-NAM. Benefits to buyers under e-NAM-be it large retailers, processors or exporters are obvious, as they can log into platform and source from any *mandi* in India connected to it. But with farmers, it may not be as simple, as most farmers do not take their crop to *mandis*; they sell off to local *arhatiya* or produce aggregator even before that. Farmers can benefit if they were to find ways for aggregating produce on their own, bypassing *arhatiya* and even local *mandi* in process. This is where FPOs/ FPCs can play a key role, by facilitating aggregation and creation of volumes that is intrinsic to success of any ambitious virtual marketplace experiment like e-NAM.

10. Agricultural Trade

India is a leading exporter of agricultural commodities. Its agricultural exports were valued at US$35 billion in 2015 (Chart 3), and its share in total world exports stood at 2.2 percent, having grown from 1.1 percent in 2000. However, annual change in India's exports of agricultural products has declined sharply from 21 percent in 2012 to (-) 19 percent in 2015[5].

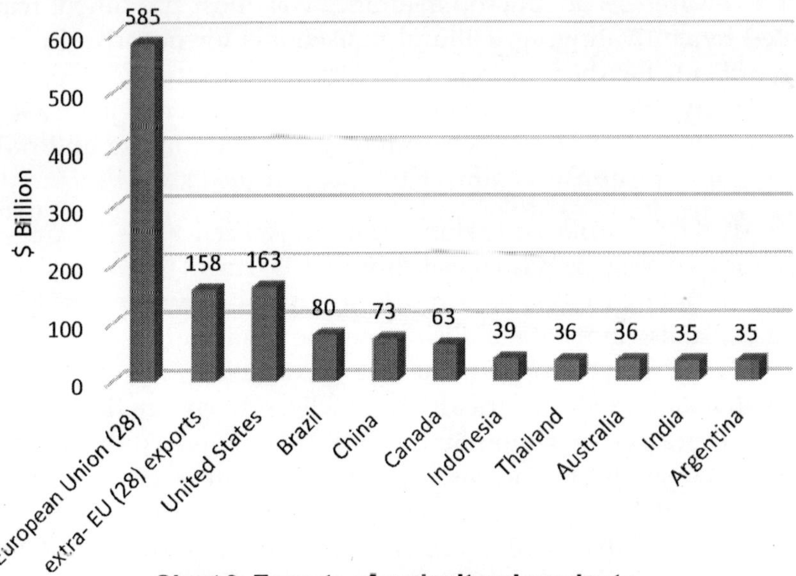

Chart 3. Exports of agricultural products (Billion USD)-Leaders Exporters-2015

Source: World Trade Statistical Review 2016, WTO.

---

[5] *World Trade Statistical Review 2016*, WTO

Small and marginal farmers lack access to export market, as well as wherewithal and knowhow to move from subsistence farming to profitable hi-tech farming for high-value produce, which could match international standards. Globalisation can have greatest impact on rural poor through their influence on agricultural sector terms of trade, availability and cost of inputs, and corporate investments in agribusiness sector through mechanisms like contract farming and corporate farming (Singh, 2016). Therefore, agricultural policy needs to focus on creating right ecosystem for opening up of export market for agricultural produce of small and marginal farmers. Integration with global markets can contribute to poverty reduction through contract farming and FPOs/FPCs.

11. Risk Mitigation

According to NSS 70th Round, a very small segment of cultivating agricultural households insured their crops against possible crop loss during reference period (July 2012-June 2013). In respect of wheat and paddy, two most harvested cereals in country, less than 5 percent of cultivating agricultural households insured their crops (GoI, 2016b). Lack of awareness about crop insurance was most prominent reason quoted by cultivating agricultural households for not insuring their crops (*ibid.*). Further, among agricultural households which had additionally insured their crops and experienced crop loss, a small percentage only received their claims. For wheat, maize and pulses non-receipt of claim was almost 100 percent (*ibid.*).

PMFBY is a path-breaking scheme which aims to provide insurance coverage and financial support to farmers in event of failure of any of notified crops as a result of natural calamities, pests and diseases. It also aims at stabilising income of farmers to ensure their continuance in farming. Further, it encourages farmers to adopt innovative and modern agricultural practices, to ensure flow of credit to agricultural sector. Union Budget 2017-18 announced that coverage of this scheme will be increased from 30 percent of cropped area in 2016-17 to 40 percent in 2017-18 and 50 percent in 2018-19. FPOs and Farmers Clubs/Federations need to educate and sensitise small and marginal farmers about importance of PMFBY in risk mitigation in agriculture, compensating for significant loss in income of farmers due to natural calamities.

Summary

In a smallholder agricultural production scenario,group approaches, viz., FPOs/FPCs, farmers' federations etc., with a focus on diversified activities, viz., agricultural crops, horticulture, dairy, fisheries, bee-keeping and off-farm sectors, could lead to profitable and sustainable agricultural production systems. This is possible through development of efficient and inclusive agri-value chains. Agricultural marketing reforms in form of e-NAM would be a win-win situation for farmers, traders, processors, retailers, and consumers. Small and marginal farmers could benefit from FPOs/FPCs, who could directly sell their produce through e-NAM network. Enabling environment could be created through massive investment in irrigation, with a focus on water-use efficiency, and value chain infrastructure, easy availability of credit (especially investment credit), and risk mitigation through wider coverage and effective implementation of PMFBY. Effective and efficient implementation of these transformative strategies, could facilitate achievement of doubling of farmers' income by 2022.

Reference

Agarwal, B. (2016). "Rethinking Agricultural Production Collectives: Case for a Group Approach to Energise Agriculture and Empower Small Farmers" in C.Ramasamy and K.R.Ashok (Eds.) *Vicissitudes of Agriculture in Fast Growing Indian Economy: Challenges, Strategies and Way Forward,* Indian Society for Agricultural Economics.

Chand, R, Prasanna, P.A. L., and Singh, A. (2011). Farm Size and Productivity: Understanding Strengths of Smallholders and Improving their Livelihoods. *Economic and Political Weekly.* 2011. XLVI (26&27).

FAO. (2015). Climate-Smart Agriculture: A call for action. Synthesis of Asia-Pacific Regional Workshop, Bangkok, Thailand, 18 to 20 June, 2015.

Foster, A. D., and Rosenweig, M.R. (2010). Is there Surplus Labor in Rural India. *Discussion Paper No. 991.* Economic Growth Centre, Yale University.

Government of India. (2016). *Economic Survey 2015-16,* Volume 2, Ministry of Finance.

Government of India. (2016a). *Economic Survey 2015-16,* Volume 1, Ministry of Finance.

Government of India. (2016b). *Some Aspects of Farming in India*, NSS 70th Round Report 573, Ministry of Statistics and Programme Implementation, NSSO, February 2016.

NABARD. (2016). Monograph on *Doubling of Farmers' Income by 2022.*

NABARD. (2016a). *Annual Report, 2015-16.*

NITI Aayog. (2016). *Report of Expert Committee on Land Leasing.*

SFAC. (2016). *Annual Report, 2015-16.*

Singh, S. (2016). "Inclusive Food Value Chains in India: Evidence and Drivers" in C. Ramasamy and K.R.Ashok (Eds.) *Vicissitudes of Agriculture in Fast Growing Indian Economy: Challenges, Strategies and Way Forward,* Indian Society for Agricultural Economics.

Swaminathan, M.S. (2010). *From Green to Evergreen Revolution-Indian Agriculture: Performance and Challenges*, Academic Foundation, New Delhi.

□□□

Chapter-20

Policy support needed for sustainable smallholder farmer production systems

C.L. LAXMIPATHI GOWDA[1] AND P. PARTHASARATHY RAO[2]

[1] Former DDG (ICRISAT) and Co-Founder, GRSV Consulting Services, Mysuru

[2] GRSV Expert Panelist and Former Principal Socio-Economist, ICRISAT, Hyderabad

---

Agriculture (that includes agro-forestry, horticulture, dairying, livestock rearing, fisheries, etc.) is backbone of any country and farming is an age-old profession of mankind. Farmers are "Annadata" for country, but they are under stress in current decade due to losses suffered in agriculture and related economic problems. The indicators of agrarian distress according Ramesh Chand (2016) are indebtedness and farmer suicides. Agrarian distress is triggered by falling farm income or sudden income loss due to crop failure or price crash. The reasons attributed to falling income or sudden income loss includes fragmentation of land holdings; natural vagaries and extremes leading to crop failures; unstable produce markets; increased cost of cultivation and overall cost of living. Thus situation of distress happenings over a period of time is a combination of several factors, which may be technological, institutional, climate related, sociological, economical, or psychological in nature.

Poverty and crop failure are not new to Indian farmer. Although farmers 4 to 5 decades ago were economically poorer compared to present day farmer, distress due to losses in agriculture and economic problems is a recent phenomenon. This agrarian distress should be an eye opener for policy makers and civil society activists at highest level to implement policies and strategies to prevent further damage to farming community.

In this paper, we look at constraints faced by smallholder farmers in India, emerging challenges for agriculture (that includes horticulture, livestock and fisheries) sector and suggest institutional and policy option to ensure SHF participation under a stable and risk-free environment.

Proliferation and Heterogeneous Nature of Smallholder Farmers

According to latest estimates, a vast majority of farmers (>75%) in South Asia, South East Asia and sub-Saharan Africa are smallholder farmers. Agriculture in Asia is characterized by smallholders cultivating small plots of land. The average size of operational holdings (actual area cultivated) is only 0.5 hectares in Bangladesh, 0.8 hectares in Nepal and Sri Lanka, 1.3 hectares in India and 3.0 hectares in Pakistan. The situation is similar in many other countries of Asia. In China 95 percent of farms are smaller than 2 hectares. In Nepal, 93 percent of operational holdings are operated by small farmers (<2 hectares) covering 69 percent of cultivated area. In Bangladesh, small farms account for 96 percent of operational holdings with a share of 69 percent of cultivated area. Pakistan is an exception, with a relatively high concentration of large landholdings (IFAD, 2009).

For purpose of this paper, we will club both smallholder and marginal farmers, and call them Smallholder Farmers (SHF) or Smallholder Producers (SHP) who own less than 2 ha per household. In 2000, SHF in India accounted for 82% of land holdings and 39% of land. In 2011, their share of small-holdings further increased to 85% and share in land to 44.5%. The average size of land holding of marginal farmers (less than 1 ha) was 0.39 ha in 2011. For smallholder farmers (1 to 2 ha) it was 1.42 ha. The average size of land holdings has been declining over time and overall at all-India level average farm size declined from 1.33 in 2001 to 1.15 ha in 2011 (Ministry of Agriculture, GOI 2015). The trend of declining land holding size would continue in near future due to factors such as growing rural population, urban growth that is not labor intensive, formal and informal barriers to rural-urban migration, and distortionary land policies. In addition to factors mentioned further fragmentation of land is due to prevailing inheritance laws in India. Though smallholder farmers account for 44% of land area they account for nearly 50% of production of food grains, fruits, vegetables etc. Many studies have also confirmed inverse relationship between farm size and productivity per hectare. Small farmers are characterized by smaller applications of capital but higher use of labour and other family-owned inputs, and a generally higher index of cropping intensity and diversification (IFAD 2009).

SHF is a heterogeneous group, with diversity in various aspects relating to land holdings, crops cultivated, degree of technology adoption, socio-economic status, cultural variations, etc. Hence, it is difficult to define SHF precisely. Possibly this is one of reasons why this group has failed to get much attention from governments for enhancing their incomes and improving livelihoods. Majority of SHF are subsistence farmers growing field and horticulture crops, and producing livestock or fisheries products mostly for household consumption. However, every farming family needs to produce some surpluses to sell and earn a little money for family needs and to buy other daily necessities. There are a few SHF who practice small-scale commercial agriculture/horticulture/livestock/fisheries and do produce reasonable quantities of marketable surpluses, and earn reasonable profits.

Constraints faced by SHF

The SHF face a number of constraints/limitations, and a few are listed below:

Fragmentation of land holdings make them uneconomical for farming. To overcome small land holdings SHF lease land from large holder farmers but generally leasing period is restricted to crop season and there is no formal lease agreement since land owner fears losing his land to leasee? Due to informal agreement SHF cannot obtain loan on leased land nor is he eligible for crop insurance. Thus tenant does not undertake any productive investments on leased land making it less productive over time due to deterioration in quality of land in terms of nutrients due to continuous and intensive cropping. Owing to lack of sufficient resources farmers do not replenish nutrient loss in soil after each season. Many times they are not even aware of nutrient loss due to lack of access to soil testing facilities.

During last few decades a number of new crop production technologies and improved varieties have been released. However, SHF have inadequate access to appropriate crop/horticulture/livestock/fisheries production technologies and inputs (seed/planting materials, fertilizers/feeds, etc.).

Smallholder farmers depend more on ground water compared to large-holding farmers who have more access on canal water and groundwater. Ground water is depleting in many areas of India. Marginal and smallholder farmers are going to face more problems for water in future. Therefore, water management is going to be crucial for these farmers (Dev. 2012). Owing to poor extension system of government, due to understaffing, farmers rely on input dealers for information seeds, fertilizer dosage, pesticides etc. Such information though useful is not best that is available.

Many SHF do not receive budgeted subsidies on fertilizers and other inputs, interest subsidy schemes on crop loans and benefits under various government schemes due leakages and lack of information on available subsidies and schemes. Non-accessibility to institutional credit and inadequate coverage of crop insurance is another major issue. SHF access to institutional credit is limited owing to small parcels of land, high transaction cost and cumbersome paper work for obtaining loans. Owing to social obligations farmers also need credit for events such as marriages, births and death in family. Many SHF obtain loans from money leaders or large holder farmers at 2-3 times higher interest rates than from institutional sources. In event of crop failure due to drought, floods or pests and diseases farmer will not be able to repay loan and in absence of safety nets is forced to sell his assets further adding to his distress. The coverage of crop insurance is very poor since insurance is linked to loans from formal sources of credit.

SHF face price risk besides production risk when prices of produce fall during harvest and also due to over production. There are a number of examples where prices crashed by more than 50% for horticulture crops like onions and tomatoes but also for coarse cereals, oilseeds, cotton etc. Although government announces Minimum Support Price (MSP), price at which government agencies will buy if market prices fall below this price, its field level implementation is poor or even not existent in remote areas in several states.

Farmers relay on middlemen or go to nearest APMC markets for sale of output. Farmers receive a lower price and do not receive premium price for better quality produce. Cartelization of traders ensures that farmers do not receive a fair price. Due to lack of storage facilities farmers are forced to sell their produce for price offered since he needs cash immediately to repay loans and meet household needs. Such sales are called distress sales as prices at time of harvest are lowest and tend to rise during off season.

SHF borrow money from commission agents/traders during crop season to meet input requirements (such as seed and fertilizers) and are forced to sell their produce to them to repay loans even if offered a lower price. Such credit and output tied transactions are common. Though this may be beneficial to SHF, as they get easy loans, they end up being exploited. For perishables like fruits and vegetables commission agents at market charge higher commission than what is stipulated as per law. SHF also face post-harvest losses due to poor infrastructure related to post harvest handling and storage for food grains but mainly for perishables like fruits and vegetables, livestock products etc.

SHF diversify their income sources as a risk mitigating strategy. Besides crops, livestock rearing, working as wage laborers and finding off-farm employment are some such avenues. However many times off-farm opportunities are limited and even if available they do not have necessary skills to take up such jobs. SHF many times end up as construction workers in urban areas and live in urban slums.

High cost of living due to changed lifestyles, without proportionate increase in income from farming results in debt and economic crisis. This may be in areas of housing, transportation, education, clothing, food and electronic gadgets. Thus SHF farmers are frequently battered psychologically due to various losses involving crop, livestock, machinery, income, land, health, and community life.

The status of smallholder farmers in developing countries, including India, has not changed much in past two decades. They continue to face many challenges, and unable to get out of poverty vortex. As someone said, Indian farmer is born in debt, lives in debt and dies in debt. This needs to change. Despite constraints, SHF in developing countries including India play a key role in meeting current and future demands for food and nutrition security for ever increasing population.

Changing context of Agriculture and Demand Driven Supply Chains

However, in past two decades, agriculture is no longer main income generating occupation for many farming communities (World Bank, 2007; Deb *et al.*, 2014). Both off-farm and non-farm activities and other enterprises are emerging as important income sources for rural communities. Majority of youth in rural areas are neither willing nor interested in residing in villages and practicing agriculture or allied activities, because they consider that agriculture is not a profitable occupation. Most of younger generation are interested in migrating to cities for jobs in mainstream services and other sectors. This trend can only be reversed or halted only when necessary policy framework and implementation are available.

Secondly, driven by structural transformation, liberalization, and globalization another important change is taking place in developing countries that have a bearing on production and profitability of all sectors including agriculture. Income growth, urbanization and change in tastes and preferences are driving demand driven commodity value chains in agriculture sector. The demand for high value commodities like fruits, vegetables, fish and livestock products is growing faster compared to food grains. To participate in demand driven value chains producers have to adhere to strict quality standards, grades and sanitary and phyto-sanitary (SPS) measures particularly for export markets. While farm sizes are getting fragmented processing sector and retail outlets are becoming bigger and bigger to reap benefits of economies of scale and to face global competition. Under this emerging scenario there are apprehensions that SHF with small scale production, and lack of knowledge of SPS standards and access to information on grades and quality specifications may be further marginalized (Parthasarathy *et al.*, 2009).

The Government of India has floated a number of schemes and development initiatives for benefit of smallholder farmers. For example, land access schemes, credit and input access schemes, promotion of cooperative societies, group farming, collective marketing, subsidies on purchase of seeds, livestock, etc. Many of schemes have benefitted a few SHF but have not changed ground level situation significantly since its targeting is an issue as also upscaling is a major challenge in increasing reach of these initiatives to needy.

Policy Support Needed for Smallholder Farmers: Easing land constraint

To ease constraint of small size of holdings, leasing of land should be legalized to extent possible such that SHF are eligible for loans and able to insure their crops on leased land. In order to overcome land owners' apprehensions on formal land leasing to tenants concept of Public Land Bank should be promoted to bridge supply and demand for land. The land bank will act as a mediator between owner and lessee and also pay a small interest to those who deposit their land papers for lease with bank. The lessee will be eligible for credit and crop insurance. In this way interest of both parties can be protected.

Efforts should be made for computerization of land records for transparency and clarity of title deeds. Land acquisition by government for any development work should be with consent of farming community who should be adequately compensated or given alternative land for farming. Although being implemented in some states, farmers should be provided soil heath cards so they are aware of quality of their lands and also lands they are leasing in.

Inputs, Credit and Subsidies

There is need to ensure that SHF have facilitated access to critical farm inputs (seed, fertilizers, chemicals, etc.) and also hassle-free process for receiving institutional credit from banks and other financial institutions at nominal rates of interest, and reduced credit interest rates for smallholders. Low-cost custom hiring of farm machinery should be promoted to enable SHF to mechanize their farming operations. Policies should promote small scale irrigation systems, rainwater harvesting in rainfed areas and participatory community irrigation management. Both government and non-government agencies should provide support to community owned and operated irrigation sources (for reasons of equity).

Policy should be changed from price subsidy on inputs to income subsidy i.e., direct transfer of funds to beneficiary accounts. It could be designed in such a way that SHF get a higher subsidy based on their land holdings. This would prevent leakage and misuse of funds and also better target subsidies towards marginal and small farmers (Gulati *et al.*, 2016).

Government should ensure that banks strictly follow lending for priority sector and within priority sector to agriculture and within agriculture set aside a fixed percentage for SHF. Governments should formulate appropriate, yet flexible, guidelines to properly classify and identify SHF for receiving support from different schemes. Otherwise, much of funds and facilities meant for SHF may be cornered by influential and politically well-connected farmers.

Risk Mitigation

Governments should provide fool-proof safety-nets and other social support systems, especially during difficult times as a result of erratic rainfall (scanty or excess rainfall), drought conditions and other climate vagaries to mitigate production and market risks. For example, support to drought proofing, improving village market infrastructure, and minimum support prices to all commodities produced by smallholder farmers, encouraging contract farming, along with support to resource pooling in rural areas.

Government should legislate simplified procedures and provide support for crop, livestock and family health insurance. Several insurance schemes like crop insurance, weather based insurance, rainfall insurance etc., have been tried but none of these have reached needy farmers. Policies should be designed to upscale insurance schemes for covering more areas and farmers. The interest subvention scheme where farmers get interest subsidy on crop loans should be phased out since funds under scheme tend to be diverted away from agriculture to other higher interest earning sources (Gulati *et al.*, 2016). Promoting universal crop insurance schemes would be more appropriate. The minimum support price should be made operational for all crops not merely for rice and wheat. Similarly price stabilization schemes should be extended to more crops.

A strong emphasis is needed for diversification of income sources in rural areas through off-farm and non-farm income generating activities. Wherever appropriate, skill development programs should be taken up so that SHF can fit into semi-skilled jobs available off-farm. Farmers should receive appropriate counselling for risk management, and financial discipline and management (also see below a detailed section on Counselling Farmers).

Participatory Technology Development

State and Central Agriculture Universities, ICAR Research Institutes/Research Centers should be geared to provide SHF-friendly and supportive technologies along with hand-holding to enable adaption and adoption of improved farming technologies using preferably participatory technology development and testing methods. Many times improved cultivars and technologies developed in research stations are not adopted by farmers since they do not meet their requirements. Capacity building programs on availability and use of new technologies should be provided through programs such as farmer field schools.

Enabling Policies and Institutions for Diversification of Agriculture

Diversifying on-farm farming systems is essential to encompass agriculture, horticulture, livestock, fisheries, sericulture, etc. Presently government policy supports production of rice and wheat and other food grain crops through various subsidies and price support. Hence growing food grains is risk free. Policies should be changed to encourage crop-diversification to high value commodities through access to technology, credit, post-harvest handling, storage and marketing of produce other than staple cereals. SAUs and ICAR institutions should provide both technology and financial help SHF to gradually shift to high-value vegetables and fruit cultivation (at least in part of their land holdings) to enhance family incomes.

New institutional arrangements should enable small holder participation in production of high value commodities. There is an immediate need to strengthen pro-smallholderfarmer output value chaindevelopment. Although a few Producer Organizations/ Producer Companies are operating, process is not being fully implemented for benefit of SHF. Government, NGOs, SAUs and ICAR institutions should promote Producer Companies by providing necessary support by way of incubation and make them succeed by hand-holding in first few years of establishment.

Promote contract farming that would enable SHF participation in demand driven value chains. Under contract farming while farmers still face production risk marketing risk is mainly taken over by contracting firm.

To overcome apprehensions that large scale processors, super markets etc., would not source produce from SHF due to small scale production that would add to their transaction cost, policies should be formulated to enable formation of farmers associations, clubs etc., for bulking and aggregation of production and sale. This would be a win-win situation for both SHF and buyers. The SHF would stand empowered in dealing with large scale buyers and buyers would get required quantities at one place while meeting their quality specification.

Post-harvest Infrastructure and Marketing

Provide dedicated advisory services for output markets (including e-marketing and real-time price information), including ware-housing linkedcredit systems. Warehouses should be constructed (possibly under Public-Private Partnership) in most of Taluks. Farmers should be given warehouse receipts for materials stored, and these should be linked to Banks who should give loans to farmers, based on warehouse receipts, to enable farmers to get money for their immediate needs. Farmers can sell their produce when prices are remunerative and repay Bank loan, and take balance money home.

The government of India brought out model marketing act with main purpose of promoting competition in marketing by allowing private sector to set up markets. The model act also provides for formation of farmers association, allowing farmers to sell produce in any market, enter into contract farming with large scale buyers, provision for bulk and direct marketing, and promote warehouse receipt system. However, since agriculture is a state subject, and although several states have amended market regulation act to include provisions of model marketing act, its ground level implementation is poor.

Promoting Youth Interest in Agriculture

It is neither realistic nor possible to retain all youth in agriculture in villages. However, it is feasible to identify youth in a family those who are more inclined to carry on agriculture and nurture these talents. Both Governments and NGOs should provide agriculture education to interested rural youth to empower young farmers to become confident and self-reliant and willing to continue with agriculture as a profession. This education or training should be mostly non-degree/diploma courses and should have a staggered schedule of classes to match with lean periods of farm operations. This will enable farmers to complete necessary farm operations, and later concentrate on learning process.

Other Policy Initiatives

Women are in forefront of many agriculture and allied activities in past two decades. However, their involvement is not being harnessed properly. We need to empower women-led rural and household enterprises for augmenting family incomes. Although government is taking several steps, and new budget has increased allocation for infrastructure, this should lead to strengthening of rural infrastructure-roads, drinking water, reliable electricity supply, mobile connectivity and internet connectivity.

Considering that most farm families have at least one mobile/ cell phone, we need to harness power of cellphone based technology for farm and community advisories on weather, rainfall, pest and disease epidemics, government schemes, market and commodity price information, employment opportunities, credit, crop insurance schemes, input availability, etc.

Need for Counselling to Farmers and Farming Communities

In earlier days of joint family system, there was social, emotional and economic cushioning provided by senior members of family in terms of advice and moral support for younger generation of farmers. With families becoming nuclear even in rural areas, this buffering capacity and moral support is no longer available. Many new/young generation of farmers take too many risks that are not supported by family, resulting in distress situation, even leading to suicides. Hence, there is need to provide appropriate counselling and advice to farmers, especially when they are in distress and have economic problems.

The benefits of counselling are to:

i) Enhancemorale and confidence level of a farmer with a positive outlook to agriculture and to life;

ii) Eradicate fear and low esteem in society about farming and being a farmer,

iii) Convince need for making agriculture a profitable and sustainable venture by adopting innovative farming methods, practices, diversified crops, and diversification of systems including horticulture, animal husbandry and fisheries activities;

iv) Educate farmers (especially youth) on need for a viable lifestyle that matches their incomes, and strict financial discipline;

v) Help farmers to make decisions on what crops/ products to produce based on trends of demand and supply and projected market price;

vi) Sow seeds of "entrepreneurship" in farmers (esp. young farmers) to take up non-farm or off-farm economic activities during free time and off season to augment family income; and provide information regarding availability of training programs for Skill/ Enterprise Development in non-farm sector;

vii) Provide necessary guidance and support to ensure retaining interested rural youth in agriculture.

It is also generally agreed that counselling should be:

a) Confidential, professional and non-judgmental short-term counselling

b) Direct telephone linkage to local services if required

c) Relevant information and referrals to services and support programs in a given area

d) Relationship coaching and practical strategies for managing difficult situations.

Counselling on Importance of Self-confidence and Human Values

Because of issues mentioned above that are responsible for farmer distress, current day farmers have lost self-confidence, and some of them have also lost willingness to live. The government and civil society organizations need to build self-confidence in these farmers, and get them back on their feet to see brighter tomorrow.

Following are suggested points to be addressed during "Farmer Counselling" programs:

- Value of human life, land and agriculture
- Agriculture is profitable, if run as a business
- Enhance self-confidence and positive attitude towards agriculture
- Concept of diversified income-generating activities for multiple sources of income
- Farm families to be engaged as entrepreneurs in non-farm, village-based enterprises
- Need for stricter financial discipline and fiscal management
- Retaining interested and committed rural youth in agriculture

Implementation of Farmer Counselling Programs

As this is a novel concept, there should be a Coordinating Agency, and a Coordinator with authority to converge agriculture, horticulture, sericulture, animal husbandry, fisheries, watersheds and rural development departments, with clearly estimated and allocated budget to successfully implement program. The following are suggested for implementation:

- Opening of Farmer Counselling Centers in each Hobli or Taluka level
- Initial Training of Trainers (1000-1500 staff) during a 2-3 day workshop
- Organizing a 2-day training of Opinion Leaders/ Lead Farmers (non-political) on counselling
- Organize counselling programs at Raitha Samparka Kendra (RSKs)

Prior to implementation, it is important to constitute an advisory group consisting of at least one sociologist, one clinical psychologist, 2-3 farm experts and 1-2 financial experts, and develop a generic protocol for counselling. This generic protocol may be used, with minor modifications, to fit context and group of farmers or individual farmer as needed.

It is important to note that success of any counselling effort will be fully achieved when this is followed up with financial counselling and appropriate crop insurance, credit and other schemes that will help poor and indebted farmers.

Summary

Agriculture is backbone for country's food and nutrition security. Smallholder producers contribute significantly to country's food security. However, smallholder farmers are in eternal debt and distress most of time. This needs to change for better, and this is possible only by concerted efforts by governments, with support from non-governmental organizations and civil society. The paper has suggested a few areas where policy support is needed and/ or to be strengthened. Agriculture should be viewed as a business and needed infrastructure and policy support provided for this long term vision. On a longer term, policies to support for smallholders need to converge towards enhancing both productivity and profitability. We should aim for less people depending solely on agriculture (but with larger holdings than available currently), through strong education and skill development programs in rural areas, especially for young farmers with special emphasis on women, for non-farm and off-farm income generating enterprises.

Reference

Ashok Gulati and Prerna Tewar. (2016). "Wake up, smell leakage," *The Indian Express*, April 11, 2016, http://indianexpress.com/article/opinion/columns/union-budget-2016-framing-sector-agricultural-credit-interest-subvention-scheme/.

Deb U, Bantilan C, and Anupama G.V. (2014). Drivers of change, dynamics of rural livelihoods and poverty in SAT India. *Research Bulletin no.26.* Patancheru-502324, India. ICRISAT. Pp:50

FAO.(2002). Small holder farmers in India: Food security and agricultural policy.*RAP publication: 2002/03.* FAO RO for Asia and Pacific, Bangkok, Thailand.

IFAD.(2009). Smallholder Farming in Transforming Economies of Asia and Pacific: Challenges and Opportunities.Discussion Paper prepared for side event organized during Thirty third session of IFAD's Governing Council, 18 February, 2009.

IFPRI Food Policy Report. (2013). From Subsistence to Profit: Transforming Smallholder Farms. IFPRI, Washington DC, USA.

MahendraDev, S. (2012). Small Farmers in India: Challenges and Opportunities. Paper presented at "Emerging Economies Research Dialogue' Beijing , China, 14-15 November, 2011.

Ministry of Agriculture, Govt. of India. (2015). *Agricultural Statistics at a Glance* New Delhi: Oxford University Press.

Parthasarathy Rao and Birthal P.S. (2009). Marketing of agricultural produce in rainfed regions of India: Emerging scenario. *Indian Journal of Agricultural Marketing*, 23(3): 125-139.

Ramesh Chand. (2016) "Farmer Needs a New Deal," *The Indian Express*, March 16, 2016. http://indianexpress.com/article/opinion/columns/farmer-needs-a-new-deal-2/

World Bank. (2007). World Development Report 2008: Agriculture for development. World Bank,Washington DC.

❑❑❑

Chapter-21

Online software for irrigation and fertigation scheduling of crops under open, green house and net house conditions with micro irrigation (SIWAAS and SAFER)

S. RAMAN

Freelance Consultant, Natural Resources Management, Mumbai

The importance of water savings and need for improvement in inputs (water and fertilizer) use efficiencies in the present context in India need no over emphasis. Realizing the importance and need for saving of water in agriculture, central and state governments have come-up with subsidy programmes periodically right from 1990s. Of all the programmes, the most recent one, known as the National Mission on Microirigation (presently merged with NMSA) had been most successful one and as on March 2013 as high as 6.5 million hectares have come under micro irrigation in country (NCPAH, personal communication) as against potentiality of 45 million hectares (Raman, 2010). Though, various stakeholders are reasonably happy with trend of micro irrigation adoption in country, yet it is a disappointing fact that most if farmers still apply excess water to their crop through micro irrigation.

A survey conducted in one of leading micro irrigation adopting states indicates that in most of crops, at farmers' fields, present water savings is around 30 to 35 percent as against potentiality of around 60 percent. One of major reasons for this excess use is lack of knowledge at farmers' level regarding actual water requirement of his crop at different stages of its growth and crop geometry and needed time of operation of designed micro irrigation system, which depends on design of system, soil type, mulched condition etc. The TASK force appointed by Government of India to suggest ways and means to popularize micro irrigation in India, in its report has indicated that if 1 million cubic meter of water is saved through micro irrigation, then government has to spend 8 Crore rupees less for development of water resources. Thus it is imperative that farmer should save as much water as possible, which will be benefitting not only farmers but also public at large as saved water can be diverted to meet demands from other sectors like domestic, industry etc. without much pressure on exchequer. Thus, there is a need to guide farmer to reduce his water use in agriculture in a more scientific way through micro irrigation and hence is development of this software.

Outline of SIWAAS (Shakti Irrigation Water Application Software)

This software works on climatological approach of estimating water requirement of crops at different stages of crop growth. Subsequently, it suggests needed time of operation of micro irrigation system depending upon design of system, soil type and mulch conditions.

Major features of software are

- It works on climatological approach.
- Based on district level meteorological data, estimates water requirement of different crops, in different locations,
- Presently it can suggest water requirement of 65 to 70 agricultural and horticultural field crops and orchard crops.
- Estimates water requirement as per crop geometry
- Estimates water requirement for different stages of crop growth
- Estimates micro irrigation system operating time, depending on lateral spacing, dripper/nozzle spacing and dripper/nozzle discharge rates.
- It can suggest operating time in different soil types viz. heavy, medium and light.
- There is provision for considering mulching aspects including type of mulches while working out operating time.

- Takes into account micro irrigation system efficiency
- It has provision for editing value of local pan evaporation so that assessment becomes more accurate.
- It has provision for entering effective rainfall value so that water requirement for particular day is assessed moreprecisely.
- It is amenable for periodical up gradations/modifications

Advantages of using software are

- Additional water savings over existing usage with drip/ sprinkler/mini-sprinkler
- More energy saving
- The government can divert additional saved energy to other sectors like industries and generate more revenue.
- Farmer will be able to irrigate more area with same system in unit time.
- More area can be brought under irrigation than with present achievement through drip
- Possibilities of less pests and diseases including weeds
- Improvement in water use efficiency resulting in more net returns per unit quantity of water used
- Based on principle of Water Saved is Water Created it is estimated that investment on Major/Medium irrigation projects for creating 1MCM of water potential is ₹ 77 lakhs. Thus with more saving of water using software there will be more savings to exchequer of government as they have to divert lesser money for development of water resources and cost this figure may be still higher now.

Types and Procedure for Registration

Basically there are two types of registrations. One is for farmers and other for research/educational institutes, commercial organizations, Nongovernmental organizations etc.

Registrations for Farmers

Any farmer who wants to make use of software can register by subscribing ₹ 150 and filling up application form. The application form can be filled up either manually or electronically or through telephone. The particulars filled up by farmer will be reconfirmed through e-mail. Hence providing an e-mail address is a must. Subsequently, based on information provided in registration form, farmer will get daily SMS as to time of operation of micro irrigation system. If farmer makes any changes subsequent to his filling up of form it has to be conveyed accordingly.

Any organization which help farmers in getting registered and subsequently help farmers throughout year, as and when required, with regard to usage of SIWAAS will be given around 15 to 20 percent of subscription amount made by farmer. The organizations may be either public like universities/KVKs etc. or NGOs or private.

Registrations for Organizations

Any organization, which is interested in using software, can have access by logging on to SIWAAS.com. The registration can be done by subscribing ₹ 3000* per district which will permit for logging for one year or for 1000 calculations whichever is earlier. After registration, subscriber will be given a password, which should be used every time for logging.

*Tentative and will depend upon number of registrations made by organization

Outline of SAFER (Shakti Fertigation)

Without proper fertigation technique, potentiality of drip cannot be achieved. Presently, drip-adopting farmers are not following fertigation technique in a scientific way. They are not applying fertilizer nutrients as per crop growth and demand. This results in poor Fertilizer Use Efficiency. Further, they are spending more money by applying, unnecessarily, imported water-soluble fertilizers instead of locally available and traditionally used water-soluble fertilizers, which are less costly. Thus there is an urgent need to guide farmers on proper fertigation technique and hence is this software.

Major features of software are

- It is based on recommendations made by State Agricultural Universities/State Line Departments like Agriculture and Horticulture on fertilizer requirement of crops.
- Considers state level/zonal level/district level recommendations as per availability.
- It has provision for adjusting fertilizer recommendations as per soil test.
- It suggests quantity of fertilizer required for whole area under crop as well as on per hectare basis.
- It suggests fertilizer requirement based on crop growth stage.
- It considers N,P and K uptake pattern for making stage-wise fertilizer application.
- Presently it can suggest fertilizer requirement of 65-70 agricultural and horticultural field crops and orchard crops.
- There is provision for considering nutrient savings through fertigation, for calculating fertilizer requirements.
- It has option for different water-soluble fertilizer combinations for different nutrients.
- It calculates N,P and K fertilizer requirements fordifferent percentages of phosphorus application as basal.
- It calculates quantity needed of N and K water-soluble fertilizers based on source of phosphorus as basal (viz. Super phosphate or DAP etc.).
- It is amenable for periodical up gradations/modifications.

Anticipated advantages of using software

- Additional fertilizer savings over existing usage with drip.
- More monetary saving as imported water soluble fertilizers are kept to minimum and less costly traditionally used water soluble fertilizers are considered.
- Improved fertilizer use efficiency as fertilizers are suggested as per crop growth stage through fertigation.

Types and Procedure for Registration

Basically there are two types of registrations. One is for farmers and other for research/educational institutes, commercial organizations, Nongovernmental organizations etc.

Registrations for Farmers

Any farmer who wants to make use of software can register by subscribing ₹ 150 per crop and fill up application form. The application form can be filled up either manually or electronically or through telephone. The particulars filled up by farmer will be reconfirmed through e-mail. Hence providing an e-mail address is a must. Subsequently, based on information provided in registration form, farmer will get periodical SMS as to quantities of different water-soluble fertilizers as per crop growth stage. Any organization which help farmers in getting registered and subsequently help farmers throughout year, as and when required, with regard to usage of SIWAAS will be given around 15 to 20 percentof subscription amount made by farmer. The organizations may be either public like universities/KVKs etc., or NGOs or private.

Registrations for Organizations

Any organization, which is interested in using software, can have access by logging on to www.fertigation.co.in. The registration can be done by subscribing ₹ 150 per district per crop, which will permit for logging for one year or for 1000 calculations whichever is earlier. After registration subscriber will be given a password, which should be used every time for logging.

Both the software is now available for open, green house and net house cultivation conditions.

□□□

Chapter-22

Small livestock farmers and the need for development-friendly production systems

RAGHAVENDRA BHATTA
Director, ICAR- NIANP, Bengaluru

About 80% farmers in developing world, including India, have small farm holdings. However, policy makers are of the opinion that small farm holders are not sustainable and that we need to find alternate livelihoods for them. On the contrary, few research institutes and innovative farmers have generated technologies and substantial evidences, indicating that carrying capacity of a given land can be enhanced significantly such that a family of five has been claimed to make a living on less than half ha area. This, however, needs inputs of assured water, knowledge, labor etc.

Why there is focus on small farmers? In developing countries such as India, agriculture continues to be the main source of employment, livelihood and income for between 50 - 90 per cent of the population. Of this percentage, small farmers make the up the majority, up to 70 - 95 per cent of the farming population. Therefore, small farmers are making significant proportion of the population. They have traditionally survived on subsistence production. Many in the last 2 decades have experimented with export crops with occasional initial success but generally with failures.

Table 1. Landholding pattern in India (2010-2011)

Category of holding	Size of group (ha)	Holding (million)	Holding (%)	Area (m ha)	Area (%)	Average size (ha)
Marginal	< 1.0	92.356	67.0	35.410	22.2	0.38
Small	1.0-2.0	24.705	17.0	35.216	22.1	1.42
Semi-medium	2.0-4.0	13.840	10.1	35.547	23.6	2.71
Medium	4.0-10.0	5.856	4.3	33.709	21.2	5.78
Large	>10.0	1.000	0.7	17.379	10.9	17.37
All		129.222	100.0	159.180	100.0	1.16

Source: GOI (2014)

In addition to the pressure put on developing countries that agricultural exports are needed to repay loans, the industrialization of agriculture was pushed by the international institutions on the premise that developing countries needed to ensure adequate food for their expanding populations. However, today, it is clear that industrial agricultural methods have not come near to resolving hunger in the world. In fact it has aggravated the food security of the most vulnerable sectors.

Industrial livestock farming in fact consumes three times more acres of land than ecologically rearing livestock. Analysts have calculated that Europe in fact uses seven times the area of Europe in Third World countries for cattle feed production. For fodder alone (including that used to produce food products for export) the Netherlands appropriates 100,000 to 140,000 square kilometres of arable land. This is often referred to as 'shadow acres' and much of it comes from developing countries. This is 5-7 times the area of agricultural land in that entire country. It has been concluded that the combination of industrial plant breeding and industrial animal breeding increases the pressure on land-use by a factor of 400% while separately increasing output of grain and milk by only a factor of 20%. One acre of a farm that only uses inputs derived from the farm itself with integrated livestock and crop production can support a family and two farm animals. If both crop production and livestock are made to rely on intensive inputs, the same family and two animals require additional shadow acres of land to produce the inputs and absorb the pollution.

Table 2. Livestock population in India

Species	Livestock population (million)		
	2003	2007	2012
Cattle	185.2	199.075	190.904
Buffalo	97.9	105.342	108.702
Sheep	61.5	71.558	65.069
Goat	124.4	140.537	135.173
Other species	16	13.164	13.185
Total	485	529.676	512.057

Source: GOI (2014)

Table 3. Milk production trend

World	
Milk production	817 Mt
Per capita availability	229gm/day
Rate of production growth	2.2 %
India	
Milk production	156 Mt
Per capita availability	337 gm/day
Rate of production growth	6.28 %

Source: IDF World Dairy Situation 2016; NDDB

Table 4. Milch animals and milk yield in India

Milch animals	Adults (million)	Share of milk (%)	Average milk production (kg/day)
Crossbreds	12.294	24.3	6.97
Native cows	32.007	20.9	2.27
Buffaloes	38.193	51.1	4.71
Goats	38.000	3.8	0.85

Source: GOI (2013)

In 1950, India was producing only 17 million tonnes. We continue to rule the global milk production market with the production increasing from 137.7 million tonnes in 2013-2014 to 156 million tonnes in 2015-2016. Dairy plays a vital role in the agricultural domain of

India and contributes about 3.9 per cent to the national gross product. However, due lack of quality germplasm, veterinary care and poor market linkage, most of the livestock have not been contributing substantial income. Unable to cope with their needs, the livestock farmers try to increase their herd size while causing pressure on fodder and feed resources and thereby contributing to global warming by emitting large amounts of green house gas in the form of Methane. The enteric methane emission from Indian livestock is to the tune of 10.0 Tg annually.

There are several hurdles for the small farmers to procure good quality feed at competitive rates. Feed constitutes more than 70 percent of the cost of the milk. No breeding programme can succeed unless the improved livestock are given adequate feed, nutritious fodder in addition to supplementation of minerals. In India most of the livestock are underfed due to the shortage of feed and fodder. While there is a severe shortage of animal feed (Table 53), however, farmers in fodder surplus areas such as Punjab and Haryana are even burning crop residues or selling them for alternate uses at substantially lower prices. To avoid this, automatic harvesters should be modified that can harvest both the grain and leave very less amount of residue on the ground. In the absence of organized supply of these inputs, the local traders take undue advantage by supplying inferior quality livestock feed at very high price. Under these situations dairy farming can become uneconomical in spite of superior inputs. The land under permanent pastures and grasslands is about 3.6 % of the geographical area and the fodder cultivation is limited to only 4.86 % of the cultivable land. It is suggested that the fodder seed production should be a part of the activity of the DAHDF instead of the Agriculture, where huge amount of money is being spent on production of various types of seeds such as breeder seed etc.

Table 5. Surplus/ Deficit of feed resources for animal feeding in India

Availability	2015			2020			2025		
(Million tonnes)	Dry Fodder	Green Fodder	Conc.	Dry Fodder	Green Fodder	Conc.	Dry Fodder	Green Fodder	Conc.
	404.1	589.81	60.53	459.8	600.53	69.62	510.2	593.8	77.78
Requirement	491.0	840.0	87.0	530.0	880.0	96.0	550.0	1000	105.0
Surplus/Deficit(%)	-17.70	-29.78	-30.43	-13.24	-31.76	-27.48	-7.24	-40.62	-25.92

Source: NIANP (2015), Bengaluru

The total requirement of livestock feed is over 200 million tonnes every year whereas the feed manufacturing companies are producing only 8 million tonnes. There is a need to increase the capacity of the feed plant and their utilization capacity by changing from two to three shift operations. Newer feed resources such as areca sheath, fruit silages developed by NIANP, Bangalore and newer feed technologies such as complete feed block, total mixed rations and by pass nutrient technologies (protein and fat) may be adopted. Therefore, the challenge is to address the problems of the marginal farmers by improving the productivity of their livestock along with a suitable cropping plan for different agro-eco zones to enhance their income. To increase the production of crop residues, there is a need to develop new verities of fodder seeds that produce more dry fodder without decreasing per acre yield of grain. The mixed farming concept has several advantages such as fodder trees grown on bund as wind breaks as well as source of fodder and efficient use of by-products such as crop residues (straws, stover and kadbi) as animal feed and utilization of dung as farm-manure. In order to ensure the success of mixed farming, it is necessary to improve the productivity of livestock and develop a sustainable land use plan depending on the soil productivity and domestic needs. As animal husbandry demands large quantity of fodder, low productive land can be used for fodder production with very remote chances of failure. This type of land use plan will increase the income from agriculture, while supporting livestock husbandry as well. The program should also include the development of efficient of efficient value chains for crop production produce and animal products.

Poor access to veterinary services, lack of efficient breeding services at the doorstep of small farmers, acute feed and fodder shortage and inadequate linkage with market for sale of produce, force small farmers to neglect their livestock and this leads to decline in their production. Therefore, even the elite breeds such as Holstein Friesian and Jersey are being neglected by the farmers. Due to sentimental reasons, social issues and legal restrictions on cow slaughter there has not been any scope for selection by culling. Most of these animals have been surviving on crop residues and free grazing on community pastures, farmers continue to maintain them as do not contribute substantial to their income. These animals serve as a liability to the farmer and pose a threat to the environment by causing pressure on bio-diversity and producing huge amount of emission of methane adding to the green house gases pool. In order to take challenge as an opportunity, good husbandry practices should be promoted, while making best use of the available land for food and fodder production for livestock.

Small farmers have the capacity to take up intensive animal husbandry to enhance their farm income. While increasing the productivity, it is also necessary to make ideal planning of their farming system to ensure that the mutual support of agriculture and livestock husbandry activities go hand in hand. For e.g. where farmers have high yielding cows and buffaloes, there is greater scope for taking up intensive fodder production instead of normal food crops. The farmer can also cultivate dual purpose crops such as sorghum, pearl millet, groundnut, maize etc. where they can harvest high grain yield while producing substantial amount of fodder for their livestock. Here, the preference should be given to crops such as baby corn and sweet corn where crops are harvested when plants are still green and rich in nutrients such as soluble carbohydrates and protein to serve as good quality fodder for productive livestock. There is ample opportunity to make best use of the cow dung through farmyard manure or vermicomposting for growing high value crops such as vegetables, flowers and fruits etc., to fetch a premium price for the produce. For the involvement of poor dairy farmers in a successful dairy husbandry program, it is necessary to address research and find solutions to their problems by Institutes like National Institute of Animal Nutrition and Physiology and National Dairy Research Institute and species specific Institutes. The major issues are lack of technical guidance to adopt good husbandry practices, poor access to breeding and health care services; poor quality animals requiring genetic up-gradation; nutritional deficiency due to shortage of feed and fodder; inefficient milk collection and marketing; poor linkages with research institute for learning new methods to increase productivity. The line departments such as the State Animal Husbandry Departments and Milk federations should take active role in this endeavour.

Mixed farming can be an ideal alternate, and farmers engaged in need to be further assisted to plan cropping based on quality of soil availability of water, market opportunities and sharing the area between fodder crops and cash crops. There is a need for developing the mixed farming system on a scientific basis. Development of fodder based farming system is another important consideration. Generally, on lands that are unproductive or under moisture stress, when chances of growing food crops are limited, farmers can be advised to grow fodder crops of short duration. In addition to economic aspect, it is also necessary to consider the availability of labour and make efficient use of this labour directly or by using suitable farm equipments or machinery. The program also needs to address efficient use of the resources, without any adverse impact on the environment. Other important considerations are the viable scale of operation which has been a major drawback for developing an efficient value chain for small farmers. As the small scale of operations presently managed by small farmers is not viable, the program should consider developing farmer producer groups (FPG) or self help groups (SHG) as a pre-requisite for successful promotion of farming by small farmers and marginal farmers. With proper interactions between the members of these groups, it is possible to develop efficient market linkages and enhance farmers' income by 25-30 percent by avoiding the middlemen. There is also scope for further increase the productivity of mixed farming by identifying suitable perennial forage species which can be established on the existing bunds, borders and inter-space between other crops. To support small farmers involved in dairy husbandry, there is a need to introduce tools for conservation of labour and energy, develop post-production system to avoid wastage and improve its nutritive value, extend the storage period to ensure year-round supply of fodder and produce fodder seeds to of improved verities to ensure easy availability at affordable price. Further, it may be very useful to link small farmers maintaining diary animals with dairy federation or civil society organization who can help them to develop their value chain and make best use of resources to enhance dairy husbandry for increasing their profitability and food security. Integrating crops and animals on the same farm allows the return of organic matter to the fields. Using ducks or geese in rice farming, for example, can reduce weeds without herbicides. These alternative methods are sometimes referred to as low-input technologies. However, this designation refers to the external imported inputs required. The amount of labour, skills and management required as inputs to make the land productive is in fact quite substantial.

❑❑❑

About the Editors

Dr S Ayyappan, has significantly contributed to the R&D of fisheries sector in particular and agriculture in general. He has served the country in different roles of Director, ICAR-CIFE, Mumbai; DDG (Fisheries), ICAR, New Delhi and Secretary, DARE and Director General, ICAR before joining as the NABARD Chair Professor at ICAR-NDRI, SRS, Bengaluru. His scientific contributions towards the developments of agriculture are well appreciated and he has been bestowed with several recognitions and awards.

Dr Letha Devi G is currently working as Senior Scientist at ICAR-NIANP, Bengaluru. Her research interests include climate change and Indian livestock and ICTs in agriculture. She has been associated with various research projects on climate vulnerability analysis of livestock farming and development of various ICT tools for livestock farmers.

Dr S Subash is working as a Scientist in SRS of ICAR-NDRI. He has been working in the areas of dairy entrepreneurship development, ICT modules development and impact analysis of dairy technologies. He has been associated with various research projects viz, dairy entrepreneurship development among SHGs, Small-holder dairy production system, milk innovation system and presently he is working in the area of documenting the status of indigenous cattle farming system of southern India.

Dr MCA Devi is working as a Senior Scientist in SRS of ICAR-NDRI. She has 26 years of professional experience in teaching, research and extension in dairy science. She has been working in the areas of dairy entrepreneurship development, ICT modules development and impact analysis of dairy technologies. She has guided a no. of PG and PhD students in dairy extension.

Dr PK Dixit is a graduate in Agriculture and Ph D in Agril. Econ. He started his career as Scientist at NDRI in 1986. Presently, Dr Dixit is working as Principal Scientist at NDRI Bangalore. The research interest of Dr Dixit envelops economic analysis of dairy enterprise, farming systems and impact studies. Dr Dixit served as Editor in Chief of Indian J. of Dairy and Biosciences.